Applied Principles of Horticultural Science

Applied Principles of Horticultural Science

Third edition

L. V. Brown
BSc (Hons), PgD (LWM), AMIAgrE, MISoilSci, Cert Ed, Csci

ELSEVIER

AMSTERDAM • BOSTON • HEIDELBERG • LONDON
NEW YORK • OXFORD • PARIS • SAN DIEGO
SAN FRANCISCO • SINGAPORE • SYDNEY • TOKYO
Butterworth-Heinemann is an imprint of Elsevier

Butterworth-Heinemann is an imprint of Elsevier Science
Linacre House, Jordan Hill, Oxford OX2 8DP, UK
30 Corporate Drive, Suit 400, Burlington, MA01803, USA

First published 1996
Reprinted 1999
Second edition published 2002
Third edition published 2008

British Library Cataloguing in Publication Data
A catalogue record for this book is available from the British Library

Library of Congress Cataloguing in Publication Data
A catalogue record for this book is available from the Library of Congress

ISBN: 978-0-7506-8702-7

For information on all Butterworth-Heinemann publications
visit our web site at http://books.elsevier.com

Typeset by Charon Tec Ltd., A Macmillan Company.
(www.macmillansolutions.com)

Printed and bound in Slovenia

Working together to grow
libraries in developing countries

www.elsevier.com | www.bookaid.org | www.sabre.org

ELSEVIER   BOOK AID
            International   Sabre Foundation

Contents

Part Three: Pest and disease

About the author

Laurie Brown is a horticultural scientist and educator. He is Director of Academex, a consultancy company aspiring to excellence in teaching and learning. Laurie is a consultant to the Quality Improvement Agency *Improvement Advisory Service.* He is an Ofsted associate Inspector and a consultant on the National Teaching and Learning Change Programme supporting specialist Subject Learning Coaches. Laurie previously worked with the Standards Unit on the design of exemplary teaching resources in the land based sector. He may be contacted for speaking and training invitations via: *info@academex.co.uk.*

Preface

Aims

This third edition is designed to enable people who wish to explore and investigate the science behind the natural resources of the land-based technologies in practical professional situations. It is intended to enable the learner to competently carry out routine scientific applications of horticultural principles.

The book is best regarded as a method to import practical understanding to the learner in a 'hands-on' environment, very much on the basis that 'an ounce of practical is worth a tonne of theory'. This is approached through action learning strategies used to relate and present land-based science as an interactive applied process. There is therefore, in one textbook, coverage of most of the primary science used in the land-based technologies.

Approach to the subject

The handbook assumes no scientific knowledge above compulsory school education. This book differs from others in that special emphasis is laid on evidence gained from practical experimental studies. Each chapter identifies the key facts learners should know. This is integrated with knowledge of the theoretical concepts gained from expanded supplementary text sections presenting a coherant stand-alone textbook.

Through observing and applying the science of horticulture the learner may increase understanding of how plants grow to become mature, healthy specimens established in either soils or containers. Practical exercises enhance study points and greater topic exploration emphasizes their relevance to growing plants, thus deepening underpinning knowledge.

The book presents exercises succinctly, in the context of their relationship within the horticultural environment. The learner should make their own notes and drawings since this practice can aid absorption and understanding in a relaxed, informal manner. The presence of questions supports independent learning and private study; they enhance the link to related studies and check that understanding has taken place. This is greatly assisted by both model answers and case studies.

Overall structure

The book is divided into three sections: Plant science, Soil science, Pest and disease.

Each section contains a number of chapters relating to a major principle of applied horticulture (e.g. Propagation; Soil water; Insects and mites).

Chapters are further partitioned into exercises which contain applied skills necessary to manipulate and manage horticultural resources (e.g. photosynthesis, flower structure, raising soil pH, fertilizers etc.). Skilful management of such resources is increasingly important in today's environment, where the difference between commercial success and failure rest largely on the speed and accuracy with which the horticulturalist completes the job. Increasingly, responsibility is being delegated to lower levels of staff and the need for skilled operatives is more apparent.

Each exercise is composed of practical tasks containing a very specific skill or competence, with which the learner engages, together with the supporting exercises, as a method to advancing their skills in the area under study.

The third edition includes more background theory to support the practical emphasis. Wherever practicable, topics are approached from the horticultural angle with the underpinning knowledge slotted in.

Each exercise clearly sets out a logical learning sequence containing:

Background	introduction to the theme and context including case studies
Aim	purpose and justification for doing the activity
Apparatus	list of resources required to complete the activity
Method	a logical sequence of data obtaining tasks
Results	tables to complete, thus ensuring all learning outcomes are addressed
Conclusions	problem-solving transfer of information gained to practical real-world environment, or new situations
Answers	answers to questions as an aid to check learning.

A complete list of all exercises can be found in the Contents pages.

Purpose

The purpose of this book is to provide a repertoire of practical investigations and student-centred learning study questions for the benefit of both instructors/teachers and learners. Greater emphasis is now placed on learning skills at the workplace. Modern apprenticeships and National Vocational Qualifications (NVQs) are examples of this and are well suited to the approach of this book.

The content have been revised to ensure compatibility with revised national curriculum from Edexcel, City & Guilds, The Royal Horticultural Society and the new 14–19 Diploma.

Manipulation of plant growth comes from a correct mastery of the interacting environmental variables (soils, organisms, plants). Through increased understanding and management of both plants and the resource of the land, the land-based technologist can substantially increase the ease of plant cultivation and profit. The investigative exercises offered in this book develop learner awareness and competence in the application of these principles to achieving this aim.

In recent years there has been rapid progress made in both teaching methods and student learning styles. Learners are better able to negotiate which lectures they attend, at what times and how long their course will be. Modularization of course structures has also enabled the learner to take only one or two modules per year and to advance at their own pace.

This handbook has been designed against this background of a more 'learner-centred' self-managed learning environment. It fully exploits the latest 'action research' that underpins active learning techniques. For example, it may be that theoretical components are covered by reference to specific books or lectures and the learner follows up by independent experimentation in the laboratory, workplace or other learning environment.

The author and publishers do not accept liability for any error or omission in the content, or for any loss, damage or other accident arising from the use of information stated herein. Inclusion of product or manufacturer's information does not constitute an endorsement and in all cases the manufacture's instructions and any relevant legislation pertaining to the use of such materials should be followed.

I lay no claim to originality. Some of the subjects have been attempted in various different ways, before. A few questions have been adapted from other books within the biological sciences. I thank those students and teachers who have written with encouragement and helpful suggestions.

L.V. Brown

Acknowledgements

I am indebted to the following people for their support and encouragement in the preparation of revisions. Steve Hewitt, City & Guilds National Proficiency Test council; Ian Merrick, Anthony Forsyhe, numerous students who forwarded helpful observations. Matthew Deans, Lucy Potter and the reviewers from Elsevier. Mike Early, for initially encouraging and motivating me to produce this publication. Dr Nigel Scopes, Bunting Biological Control Ltd, for comments on Biological Control exercises. Carol Oldnow, for some technical data.

Slides of biological control organisms were kindly supplied by the following: Sue Jupe, Public Relations and Marketing Services, Defenders Ltd; Dr Mike Copland, Wye College and Peter Squires, Senior Entomologist Koppert Biological Systems. Rebecca Brown for helping with slug control. Finally, my parents, Sid and the late Audrey Brown, for providing a suitable environment to enable me to work undisturbed on the initial manuscript.

Part One

Plant science

Chapter 1   Plant kingdom: classification and nomenclature

Key facts

1. Lower plants reproduce by spores, higher plants reproduce by seed.
2. Classification is used to help us make sense of the natural world.
3. The binomial system describes how to correctly use plant names.
4. Flowering plants are 'higher' plants.
5. There are two types of flowering pants, monocotyledons (one seed leaf on germination) and dicotyledons (two seed leaves on germination).

Living organisms may be simply grouped into two kingdoms: plants or animals. Plants are the major group of living organisms that have their cells surrounded by cell walls built of fibres. They are also the group of primary importance in horticulture.

All the higher plants and most lower plants, including green algae, contain pigments such as chlorophyll that can trap solar radiation energy, to make energy-rich carbohydrates by photosynthesis. The non-green plants such as bacteria and fungi do not normally contain chlorophyll, and have to obtain their energy-rich compounds from other organisms. This type of organism is called a parasite.

The plant kingdom is extremely large – estimated at three hundred thousand different species – and if fungi are included this rises to an estimate of four hundred and fifty thousand. This includes two hundred and fifty thousand species of flowering plants alone.

Six kingdom division

It is helpful to classify plants into similar groups and to name plants with a name that is understood all over the world. Plants are classified into **lower** and **higher** plants. Lower plants reproduce by spores. Higher plants reproduce by seed (see Figures 1.1 and 1.2). Using this knowledge, six simplified divisions of the plant kingdom offer sufficient detail to correlate plants with their significance in horticulture (see Table 1.1).

Figure 1.1 Higher plants reproduce by seed, like this dandelion

Figure 1.2 Lower plants reproduce from spores, like this fern

Some scientists observe a staircase of complexity, advancing from algae, through fungi, lichen, mosses, ferns and horsetails to the flowering higher plants. While the external structures of roots, shoot, stem, leaves, flowers, fruits and seed are clearly different, the internal biochemical reaction of photosynthesis is uniformly complex. Chloropyll is present even in algae, enabling photosynthesis (see Chapter 5). Photosynthesis is the most complex part of a plant's metabolism and is evidence of

Table 1.1 Plant kingdom divisions

Division	Significance in horticulture
Lower plants (reproduce by spores)	
Chlorophyta (algae)	• Very simple plants • Indicator of damp conditions • Indicator of nitrogen and cause pollution in ponds and streams • Sand filters are used to remove algae from irrigation waters
Mycophyta (fungi)	• Very simple plants containing groups of cells in threads • Recycle organic matter and nutrients • Source of disease and parasites
Lichenes (lichens)	• Very simple plants • A symbiotic relationship between algae and fungi • They are indicators of clean air/pollution, often used for decorative effect (rural planning authorities have been known to insist that rebuilt church walls are smeared with a mixture of yoghurt and dung, to stimulate lichen growth!)
Bryophyta (mosses and liverworts)	• Source of peat/compost for container grown plants • Decorative use • Smother out grass • Indicates dampness • Sometimes called the amphibians of the plant world
Pteridophyta (ferns and horsetails)	• Plants which have stems, roots and leaves • Ferns make pots and garden specimens and like shady position • Horsetails are problem weeds not responding to weedkillers
Higher plants (reproduce by seed)	
Spermatophyta (seed-bearing)	• The most important group in horticulture. Plants with stems, roots, leaves and a transport system. They may be either gymnosperms (cone-bearing, the seed is born 'naked' e.g. conifers) or angiosperms (flowering plants with seed within a fruit either monocotyledonous or dicotyledonous)

Figure 1.3 Algae

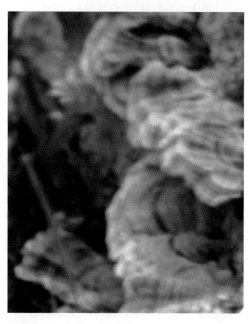

Figure 1.4 Fungi

Figure 1.5 Lichen

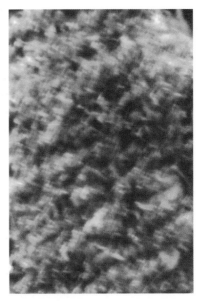

Figure 1.6 Moss

'irreducible complexity'. This means there are many ingredients needed at the same time to make the process work and all those ingredients are present and operating in the lower plants as well as the higher plants.

The wide variety of plant types found represent different survival pressures in their natural environment (e.g. cacti from desert lands, willow from wet lands). Plants can be grouped fairly easily by characteristics or modes of life. Such a classification system enables us to identify any plant from any country. It may also enable predictions about the behaviour of unknown plants from knowledge of similar plants from the same group (e.g. *Rosaceae* family are all susceptible to fireblight disease).

Figure 1.7 Ferns

Figure 1.8 Flowering plants (angiosperms)

Figure 1.9 Conifers (gymnosperms)

Alternative kingdom division classifications

Over 1.5 million species of organism have been identified on earth. Any system adopted remains only one possible interpretation of facts and observation and many different systems have developed, taking different characteristics from which a classification has developed. The history of this is presented in Table 1.2.

Table 1.2 Development of classification systems

Two-kingdom classification (animal and plant)

Animals absorb, digest food and move around. They derive their energy from other organic material and are called **heterotrophic**.

Plants, although they do have limited movements, remain stationary. They manufacture food by photosynthesis and are called **autotrophic**.

Three-kingdom classification (plants, animals and fungi)

Fungi don't fit either group. They are non-green, stationary plants that obtain their energy by parasitizing other organisms.

Four-kingdom classification (plants, animals, fungi and protista)

Protista are single-celled organisms.

Five-kingdom classification (plants, animals, fungi, protista, monera)

Blue-green algae were reclassified as cyano-bacteria which, with bacteria, became called **monera**, meaning single/solitary.

Additional two-kingdom classification (eukaryotic and prokaryotic)

Eukaryotic organisms are those with cells containing a separate nucleus. **Prokaryotic** organisms are those with no clear separate cell nucleus.

Viruses still do not fit any classification. New studies have also revealed archaebacteria which are biochemically very different from others and may need a new grouping.

Examples of the use of the classification system world wide are given in Table 1.3. Note particularly the word endings (eg. 'aceae' for family), which are standardized and also that genus and species appear in *italics* (or alternatively, may be underlined).

Figure 1.10 demonstrates how a classification map is constructed and provides an overview of the main points in classification and

Table 1.3 Classification system usage

	Example usage	
	Scientific terms	Common terms
Kingdom	Plant**ae**	Plants
Division	Spermato**phyta**	Seed-bearing plants
Class	Angiosperm**ae**	Flowering plants
(subclass)	Dicotyledon**ae**	Dicotyledon
Order	Ran**ales**	Ran**ales**
Family	Ranuncul**aceae**	Ranuncul**aceae**
Genus	*Ranunculus*	*Ranunculus*
Species	*repens*	*repens*
Common name	Creeping buttercup	Creeping buttercup

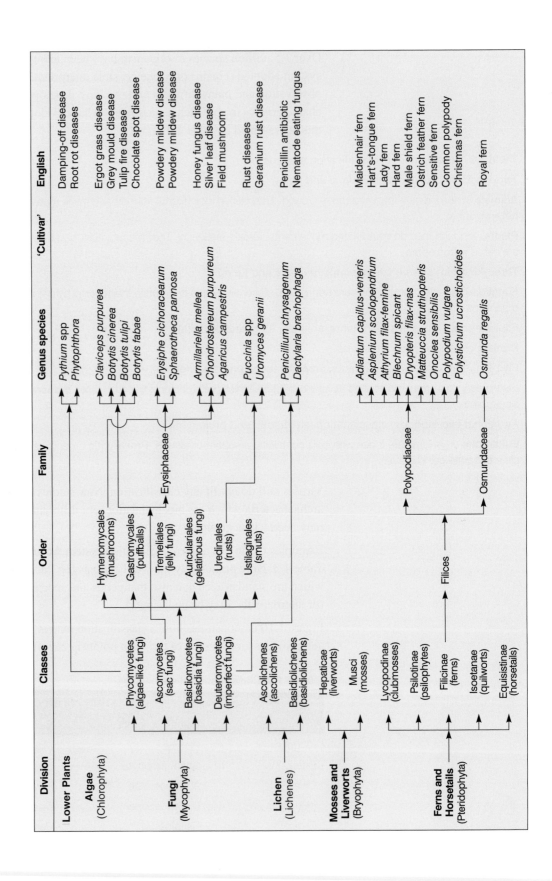

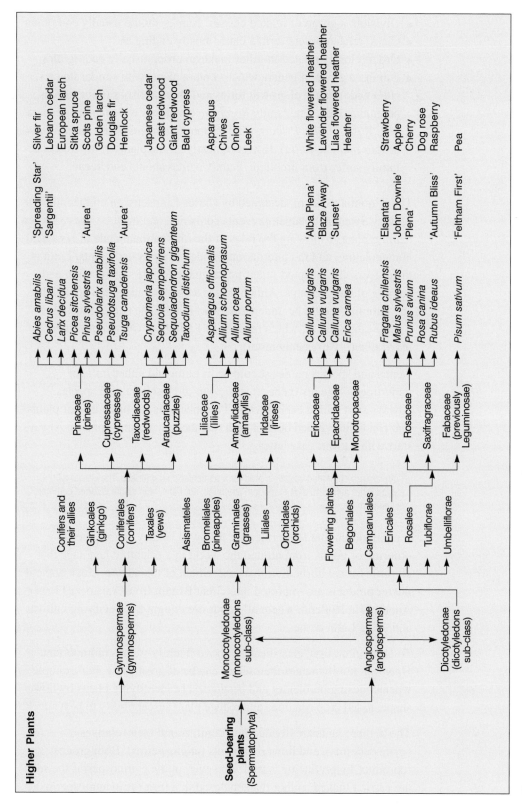

Figure 1.10 Simplified classification of the plant kingdom showing some example pathways

nomenclature. These areas are subject to opinion and consensus with no overall authority:

- Division: a group of related classes. Names should usually end **phyta**
- Class: related orders form a class, mostly ending **ae**
- Order: closely related families make an order, usually ending **ales**
- Family: a group of genera which contain plants with similar flowers, but a wide variety of growth forms and characteristics (trees, shrubs, perennials etc.). The family *Rosaceae* contains a wide range of genera (e.g. *Rosa, Malus, Prunus, Cotoneaster, Crataegus, Pyracantha, Rubus, Sorbus and Potentilla*). However, nearly all the plants in the Rosaceae family suffer from fireblight. Family names should end **aceae**.

The Binomial System, designed by Carlos Linnaeus, an eighteenth-century Swedish botanist, prevents confusion and must be obeyed when naming plants following the rules of the international code of botanical nomenclature, and the international code for cultivated plants. Latin is the universally accepted scientific language for the specification and naming of these groups. Although it is a dead language not spoken by any nation it should be understood worldwide. The names must have a genus and species, which must be in Latin (e.g. *Chrysanthemum morifolium* – Chrysanthemum).

Genus and species names should be in Latin and either *italic* or underlined. A genus (genera) is a group of closely related species. Always start with a capital letter. The species is the basic unit of plants that can be interbreed but not with members of another species. Always start with a lower case letter.

> Cultivars are cultivated varieties, bred for a particular characteristic. They begin with a capital letter and should be in modern language, not Latin, with single inverted commas ' ' or the term **cv**.

It is important that these rules are followed as they enable speedy ordering of goods from around the world in a scientific language that all professional horticulturalists use. Much of the nursery stock and cut flower products are imported into Great Britain from Aalsmeer Flower Auctions in Holland, where all goods are bought by specifying cultivar and using Latin names.

To horticulturalists, classification above 'Family' seems unimportant. However, family names themselves can be of great value. For example when choosing herbicides and pesticides; or identifying plants by flower shape, to aid crop rotations and reduce plant susceptibility to pest attack.

The higher plants are divided into conifers and their relatives (gymnosperms), and flowering plants (angiosperms). Both groups reproduce by producing a seed. However, in the gymnosperms the seed are carried 'naked' rather than enclosed in a fruit (as in angiosperms). The seeds are often borne into cones, but may sometimes be surrounded by fleshy outgrowths from the stem. For example *Taxus baccata* – the yew. Included in this group are conifer relatives, such as *Ginkgo biloba*, found as a 'living fossil'.

Exercise 1.1

The plant kingdom

Background

The plant kingdom may be conveniently divided up into groupings of plants as they affect horticulture. The characteristics and patterns of these divisions have been described in this chapter.

Aim

To investigate the division of the plant kingdom and its effect on horticulture.

Apparatus

Introductory text

List of terms: algae angiosperm; dicotyledons; ferns and horsetails; fungi gymnosperm; higher plants; lichen; lower plants; monocotyledons; mosses and liverworts.

Useful websites

www.scienceyear.com/outthere/lifemasks/plantae.php

www.plant-talk.org/93.htm

www.bgci.org/education/1776/

Method

Match the correct terminology to the sentence describing the feature.

Results

Write the correct term next to each statement.

1. Plants that reproduce by seed.
2. These plants reproduce by spores.
3. Seed-bearing plants with naked seed.
4. These organisms have their cells arranged in threads.
5. Plants with stems, roots and leaves but not transport vessels; often weeds.
6. Plants with seeds having one seed leaf.
7. A division mostly used in decorative horticulture.
8. A group of plans mostly used as a growing media.
9. The presence of these plants in irrigation water may indicate pollution.
10. Seed-bearing plants with seed contained in a fruit.
11. Plants with seeds having two seed leaves.

Conclusions

1. State the common name for the following divisions of the plant kingdom:
   Bryophyta
   Chlorophyta
   Lichenes
   Mycophyta
   Pteridophyta
   Spermatophyta

2. Describe the plant characteristics botanists use to classify lower and higher plants.
3. Why are ferns slightly more adapted to life on land than mosses and liverworts?
4. State two situations where lower plants are used in horticulture.

Exercise 1.2

Principles of classification

Background

The divisions of the plant kingdom are divided into plants with similar characteristics. They are classified into units of increasing specialism from 'Division' right down to 'Cultivar'. There are rules for the way to write down plant names, including word endings, (such as the Binomial System) and this may help to aid recognition of plant characteristics with which you may be unfamiliar.

Aim

To investigate some horticultural plants as a method to increase understanding of classification.

Apparatus

Introductory text
List of terms found with each statement

Useful websites

www.linnean.org/index.php?id=298

www.nhm.ac.uk/nature-online/science-of-natural-history/taxonomy-systematics/fathomseminar-whatsinaname/session1/no-science-tax-fathomseminar-whatsinaname-session1.html

Method

Read and follow the instructions for each of the statements.

Results

Enter your result in the table provided.

1. Number the following terms into logical structured order:

Kingdom Order Class Species Sub class Family Genus Division Cultivar

2. In the following four plants state which parts of their names are:

	Apple	Bean	Hollyhock	Annual meadow grass
(a) Kingdom				
(b) Division				
(c) Class				
(d) Order				
(e) Family				
(f) Genus				
(g) Species				
(h) Cultivar				

Apple: Plantae Spermatophyta Dicotyledonae Rosales Rosaceae *Malus domestica* 'Golden Delicious'

Bean: Plantae Spermatophyta Dicotyledonae Rosales Fabaceae *Phaseolus vulgaris* 'Stringless Green Pod'

Hollyhock: Plantae Spermatophyta Dicotyledonae Malvales Malvaceae <u>Altheae rosea</u> 'Newport Pink'

Annual meadow grass: Plantae Spermatophyta Monocotyledonae Graminales Poaceae <u>Poa annua</u>

3. With reference to Figure 1.10 construct a classification map for the following plants:
   (a) dog rose
   (b) pea.

4. Name endings are often useful to help identify the classification of unfamiliar plants. State which of the following terms are: (a) family, (b) order:

Coniferales	Rosaceae	Begoniaceae	Taxales	Poaceae
Orchidales	Magnoliales	Urticales	Cannabaceae	Ericaceae
Ericales	Aceraceae	Solanaceae	Asteraceae	Cupressaceae
Pinaceae	Graminales	Taxodiaceae	Lilaceae	Palmaceae
Urticaceae				

5. There are nearly 100 separate Family names. Observe the following list of plant families:

Asteraceae	(previously Compositae)	daisy family
Brassicaceae	(previously Cruciferae)	wallflower family, including many vegetables
Hyperiacea	(previously Guttiferae)	St John's wort family
Lamiaceae	(previously Labiatae)	mint family, often aromatic
Apiaceae	(previously Umbelliferae)	parsley family, often aromatic
Fabaceae	(previously Leguminosae)	peas and beans family
Poaceae	(previously Graminae)	grass family

Using your knowledge of classification and nomenclature:

(a) explain the rules that made it necessary to rename these seven families
(b) explain why some modern books may be in error by continuing to use old family names.

Conclusions
1. Describe the importance of Family names in horticulture.
2. State, giving one example in each case, the correct word endings for:
   (a) Division
   (b) Class
   (c) Order
   (d) Family.
3. State two ways in which plant classification helps horticulturists.

Exercise 1.3

Nomenclature for plant ordering

Background

The binomial system sets out the rules by which plants should be named. The effect of this in horticulture is upon plant ordering in what is now a global market, whether it be importing cut flowers from Holland, semi-mature trees from Italy, chrysanthemum cuttings from the Canary Islands, Christmas poinsettias from America, Cape fruit from South Africa, vegetables from Kenya, or hedging plants from the local garden centre. The internationally recognized language is Latin. It is helpful, but not essential, to specify the Family name. Genus and species should always be stated although sometimes ordering by cultivar is sufficient (but is technically incorrect). This is often done when we know precisely the effect a cultivar would give, such as when ordering cut flowers.

e.g. *Alstroemeriaceae Alstroemeria aurantiaca*

'Capri' – pale pink flowers

'Stayelar' – striped pattern flowers

'Zebra', 'Orchid', 'Canaria' – orchid-type flowers

'Rosario', 'Jacqueline' – butterfly-type flowers

Aim

To gain confidence in the use of plant nomenclature to order plants accurately.

Apparatus

   Introductory text

List of terms appearing in each section. Figure 1.10

Useful websites

www.linnean.org

www.flowercouncil.org/uk

Method

Follow the instructions given in each of the statements below.

Results

Enter your observations in the table provided.

1. For each of the following plant names state which refers to:

	Scarlet pimpernel	Transvaal daisy	Yew	Snap dragon	Gilly flower
(a) the species					
(b) the family					
(c) the genus					
(d) the cultivar					
Primulacaeae		*Anagallis arvensis* 'Golden Yellow'		Scarlet Pimpernel	
Asteraceae		Gerbera jamesonii		Transvaal Daisy	
Ericaceae		*Calluna vulgaris* 'Alba Plena'		summer white flowering heather	
Taxodiaceae		*Taxus baccata* 'fastigiata'		Yew	

	Scarlet pimpernel	Transvaal daisy	Yew	Snap dragon	Gilly flower
Scrophulariaceae		<u>Antirrhinum majus</u> 'His Excellency'		Snap dragon	
Mathiola incana		'Francesca'		red flowered Gilly flower	

2. Which **one** of the following orders is correctly written for Antirrhinums (snap dragon)?

(a)	Scrophulariaceae	*Antirrhinum majus* "His Excellency"	(intermediate flowers)
(b)	*Antirrhinum Majus*	"His Excellency"	(intermediate flowers)
(c)	Scrophulariaceae	<u>Antirrhinum majus</u> 'Coronette'	(tall flowers)
(d)	Scrophulariaceae	antirrhinum majus 'little darling'	(dwarf flowers)
(e)	*Antirrhinum* majus	'Coronette'	(tall flowers)
(f)	*Antirrhinum majus*	'his Excellency'	(intermediate flowers)

3. State what is wrong, if anything, with each of the following seed orders
   (a) Poaceae *Alopecurus pratensis* 'Aureus' (golden foxtail) (an ornamental grass)
   (b) Alopecurus pratensis 'Aureus'
   (c) Alopecurus pratensis "Aureus"
   (d) *Alopecurus pratensis* golden foxtail
   (e) <u>Poa triviales</u> (a lawn grass)
   (f) Rutaceae *Citrus Limonia* (lemon)
   (g) *Citrus Limonia*
   (h) *Citrus limonia*
   (i) Solanaceae <u>Lycopersion esculentum</u> (tomato)
   (j) Solanaceae <u>lycopersion esculentum</u>

4. The following order was received by a local garden centre to supply immediate summer colour to a small garden project. Circle all **six** technical errors.
   (a) Transvaal Daisy (Gerberas): Asteraceae *Gerbera Jamesonii*
   (b) Gilly Flowers:

<u>*Mathiola incana*</u>	"Francesca"	(red flowers)
<u>*Mathiola incana*</u>	'arabella'	(lavender flowers)
	'Debora'	(purple flowers)

   (c) Summer flowering heathers:

Ericaceae Calluna vulgaris	'Alba Plena'	(white flowers)
Calluna vulgaris	'Alba Rigida'	(white flowers)
Calluna Vulgaris	'Allegro'	(purplish-red flowers)

Conclusions

1. State the meaning of the term 'species'.
2. State the correct word endings for 'Family' names.
3. In which language should the genus and species names be written?
4. What variations should be made between writing genus and species?
5. Explain the correct way to order by cultivar.

Exercise 1.4

Flowering plants (angiosperms)

Background

The flowering plants (angiosperms) are divided into the two sub-classes called monocotyledon and dicotyledons and may be visually identified quite easily using the following guidance.

Figure 1.11 Flowering plants (tulips)

Flowering plants

The focus of most aspects of horticulture revolves around the growth and development of flowering plants (angiosperms). These always have developing seeds enclosed in a fruit and flowers, highly adapted for insect or wind pollination. They are divided into two sub-classes: the **monocotyledons** and the **dicotyledons** (some times abbreviated to monocot and dicot, respectively). They have many differences which can effect our management of these plants (e.g. it is difficult to take cuttings from monocotyledons).

Figure 1.12 Monocotyledons have parallel leaf veins.

Figure 1.13 Dicotyledons have branching leaf veins

Dicotyledons

'Di' means two, 'cotyledon' means seed leaf

These are plants bearing two cotyledons or seed leaves in the seed. (e.g. broad leaf trees, roses, daisies, cabbages, pelargoniums, cacti, oak, tomato).

All seeds contain a juvenile plant embryo and a food reserve normally as a cotyledon. The cotyledon supplies energy until the germinating plant is able to photosynthesize.

The seed leaves or cotyledons may emerge above the ground with the shoot. The seed leaves turn green and can be confused with the true leaves still to

Figure 1.14 Monocotyledons have floral parts in multiples of threes like the 6 petals of this Amaryllis

Figure 1.15 Dictoyledons have floral parts in mltiples of fours or fives

expand from the shoot. Sometimes they remain below the ground, supplying their food reserve to the developing seedling.

Monocotyledons

> 'Mono' means one, cotyledon means seed leaf

This refers to a plant bearing only one cotyledon or seed leaf when they germinate (e.g. grass, lilies, orchids, palms, bromeliads, daffodils, crocus, palm tree, spider plant, pineapple, wheat, barley, oats, iris, onions, gladiola, tulip, narcissus).

They have parallel veins rather than the net-like veination of the dicotyledons.

The flowers of the monocotyledons are in threes or in multiples of three, as opposed to dicotyledons with their flower parts in multiples of fours or fives. It is very rare for monocotyledons to form shrubs or trees because they don't form a cambium ring (growth tissue) to allow stem thickening.

The essential differences between monocotyledons and dicotyledons are summarized in Table 1.4.

Table 1.4 Distinguishing features between monocotyledonous and dicotyledonous plants

Dicotyledon	Monocotyledon
Embryo seed **(Exercise 2.1)**	
2 seed leaves (cotyledons)	1 seed leaf (cotyledon)
Roots **(Exercise 4.3)**	
Primary root often persists and becomes strong tap root with smaller secondary roots; cambium present	Primary roots of short duration, replaced by adventitious root system. No cambium
Xylem in a star shape	Xylem in a ring

Growth form	
herbaceous or woody	mostly herbs, few woody
Pollen (Exercise 9.4)	
Has 3 furrows or pores	Has 1 furrow or pore
Stem vascular bundles (Exercises 8.1 and 8.2)	
Arranged in a ring round the stem; cambium present; secondary growth in stem; stem differentiated into cortex and stele	Scattered around stem; no cambium; no separation into cortex and stele
Leaves (Exercise 5.1)	
Net veined; often broad; no sheathing at base; petioles (stalk) present	Parallel veins; oblong in shape; sheathing at base; no Petiole (stalk)
Flowers (Exercises 9.1, 9.2 and 9.3)	
Parts usually in multiples 4 or 5	Parts usually in multiples of 3

Aim

To identify monocotyledon and dicotyledon sub classes based upon visual observation.

Apparatus

Previous text section, Table 1.3

Cabbage	Lily
Chrysanthemum	Pelargonium
Crocus	Rose
Daffodil	Spider plant
Geranium	

Useful websites

www.csdl.tamu.edu/FLORA/201Manhart/mono.vs.di/monosvsdi.html

www.emc.maricopa.edu/faculty/farabee/biobk/BioBookPLANTANATII.html

www.nccpg.com/Page.Aspx?Page=440

Method

Describe the specimen plants' characteristics under the headings given below.

Results

Enter your observations in the table provided.

Description of plant tissue in a range of plants

Plant tissue being described	Cabbage	crocus	geranium	lily	pelargonium	Rose	spider plant
Leaves (net veins or parallel veins?)							
Flowers (petals in multiples of 4/5 or 3s?)							
Roots (tap or fibrous?)							
Growth form (herbaceous or woody?)							
Deduction: monocotyledon or dicotyledon?							

Observe Plant A and Plant B. State, giving reasons, whether each is a member of the sub-class monocotyledon or dicotyledon.

Specimen	Sub-class	Reasons
Plant A, daffodil		
Plant B, chrysanthemum		

Conclusions

1. Explain the meaning of the terms 'monocotyledon' and 'dicotyledon'.
2. List four differences between monocotyledons and dicotyledons and name two examples of each.
3. State one significance in horticulture of the practical difference between the two sub-classes.

Answers

Exercise 1.1. The plant kingdom

1.   higher plants
2.   lower plants
3.   gymnosperm
4.   fungi
5.   lichen
6.   monocotyledons
7.   ferns and horsetails
8.   mosses
9.   algae

10.   angiosperm
11.   dicotyledons

Conclusions

1.   See Table 1.1.
2.   Method of reproduction. Higher plants reproduce by seed. Lower plants by spores.
3.   Presence of root system enables more efficient water and nutrient use.
4.   Decomposed moss is the major growth medium used in horticulture. Ferns are used for decorative effect. Other uses may be deduced from Table 1.1.

Exercise 1.2. Principles of classification

1.   kingdom (1), order (5), class (3), species (8), sub-class (4), family (6), genus (7), division (2) cultivar (9).

2.   **Apple**: (a) Plantae, (b) Spermatophyta, (c) Dicotyledonae, (d) Rosales, (e) Rosaceae, (f) *Malus*, (g) *domestica*, (h)'Golden Delicious'.

    Bean: (a) Plantae, (b) Spermatophyta, (c) Dicotyledonae, (d) Rosales, (e) Fabaceae, (f) *Phaseolus*, (g) *vulgaris*, (h)'Stringless Green Pod'.

    Hollyhock: (a) Plantae, (b) Spermatophyta, (c) Dicotyledonae, (d)Malvales, (e) Malvaceae, (f) <u>Altheae</u>, (g) <u>rosea</u> 'Newport Pink'.

    Annual meadow grass: (a) Plantae, (b) Spermatophyta, (c) Monocotyledonae, (d) Graminales, (e) Poaceae, (f) <u>Poa</u>, (g) <u>annua</u>

3.   (a) spermatophyta angiospermae Rosales Rosaceae *Rosa canina* (dog rose).
      (b) spermatophyta angiospermae Rosales Fabaceae *Pisum sativum* 'Feltham First' (pea).

4.   (a) Rosaceae, Begoniaceae, Poaceae, Cannabaceae, Ericaceae, Aceraceae, Solanaceae, Asteracea, Cupressaceae, Pinaceae, Taxodiaceae, Lilaceae, Palmaceae, Urticaceae.

      (b) Coniferales, Taxales, Orchidales, Magnoliales, Urticales, Ericales, Graminales.

5.   (a) The old family names did not end 'aceae' and therefore did not meet the international conventions. They were renamed with appropriate names to correct this.
      (b) The old names used in some text books are out of date and the practice should be avoided. However, editors might argue that they are a 'synonym'. A name in common use and preferred, or misaplied, to the correct name.

Conclusions

1.   Family names contain plants with similar flowers but a wide variety of growth forms. Knowing the family helps with identification of unknown plants, e.g. Lamiaceae contains many plants with square stems. Knowledge can also be applied to exploit the differential toxciticy of weed killers.

2.   (a) phyta, e.g. spermatophyta (seed-bearing)
      (b) ae, e.g. angiospermae
      (c) ales, e.g. Ranales
      (d) aceae, e.g. Ranunculaceae

Exercise 1.3. Nomenclature for plant ordering

1. Scarlet Pimpernel: (b) Primulacaeae, (c) *Anagallis,*
    (a) arvensis, (d)'Golden Yellow'.
    Transvaal Daisy: (b) Asteraceae, (c) <u>Gerbera</u>,
    (a) <u>jamesonii</u>.
    Yew: (b) Taxodiaceae, (c) *Taxus,* (a) *baccata,*
    (d) 'fastigiata'.
    Snap dragon: (b) Scrophulariaceae, (c) <u>Antirrhinum</u>,
    (a) <u>majus</u>, (d) 'His Excellency'.
    Gilly flower: (c) *Mathiola,* (a) *incana,* (d) 'Francesca',
    (b, absent).
    Heather: (b) Ericaceae, (c) *Calluna,* (a) *vulgaris,*
    (d) 'Alba Plena'.

2. (c) All others either have the genus species or
         cultivar incorectly written.

3. (a, b, e, h, and i) are all correct.
    (c) Cultivar should be in single inverted commas.
    (d) Cultivar not identified by capital letter or
         inverted commas.
    (f) Species should be written with a lower case l.
    (g) Species should be written with a lower case l.
    (j) Genus should start with a capital L.

4. (a) *Jamesonii* is the species and should start with a
         lower case letter.

(b) Genus and species are both italic and
     underlined. 'Francesca' is the cultivar and should
     be in single inverted commas. Arabella should
     start with a capital letter.
(c) Calluna vulgaris should be in italics or
     underlined. Vulgaris is the species and should
     start with a lower case letter.

Conclusions

1. The species is the basic unit of plant classification
    containing plants that can interbreed but not with
    members of another species.

2. aceae.

3. Latin.

4. Genus always start with a capital letter, species with
    a lower case letter.

5. The genus and species should be stated first then
    the cultivar. Subsequent cultivars from the same
    species can be stated without further reference to
    genus or species.

Exercise 1.4. Flowering plants

Dicotyledonous plants; cabbage, geranium,
pelargonium, rose

Monocotyledonous plants: crocus, lily, spider plant

Plant A (Daffodil): Monocoyledon. Leaves with parallel
veins. Flower parts in multiples of 3

Plant B (Chryanthemum): Dicotyledon. Leaves with net
veins. Flower parts in multiples of 4/5's

Conclusions

1. Monocotyledon = one seed leaf.
    Dicotyledon = two seed leaves.

2. For differences see Table 1.3. Example monocots
    include lily, spider plant. Example dicots include
    geranium and chrysanthemum.

3. Absence of a cambium in monocotyledons makes
    taking cuttings very difficult. Other propagation
    techniques including micropropagation may be used.

Chapter 2 Seed propagation (viability and vigour)

Key facts

1. Vigor is the ability to germinate under different environmental conditions.
2. Viability is percentage germination rate predicted.
3. Dormancy is a state when there is no visible sign of growth.
4. Techniques to break dormancy include priming, scarification, stratification, and acidification.
5. Epigeal germination – the seed leaves appear above ground on the seedling.
6. Hypogeal germination – the seed leaves remain below ground in the seed.

Background

Flowering plants begin their life cycle as seeds. The seed contains a miniature plant called an embryo, from which a new adult plant will grow. The young plant that develops from a germinating seed is called a seedling.

Seed quality (viability and vigour)

Most edible horticultural crops and bedding plants are grown from seed (see Chapter 3). In selecting seed to grow, quality becomes an important aspect. Our aim is to produce synchronized germination, uniform crop characteristics (e.g. height) and healthy, robust growth.

There are several properties that are used to help assess the quality of seed such as

- purity: percentage of undamaged seeds
- germination: percentage of viable seed
- health: presence of potential pathogens.

For edible horticultural crops and forest trees, the **International Seed Testing Association** governs the test procedure for these three characteristics, under the authority of **The Seed Acts**. A minimum germination percentage must be achieved. Flower seeds are not covered by the Act. Germination percentage (**viability**), will vary under different environmental conditions. For example, more seed is likely to germinate in sheltered lowlands than in the cold high altitude of mountains and there is considerable variability in performance. This variation in performance is called **vigour** and this property is regulated by the **Vigour Test Committee** of The International Seed Testing Association. In North America the **Vigour Committee of the Association of Official Seed Analysts of North America** perform a similar role. Our interest in vigour lies in knowledge of the proportion of seedlings established in the field and the rate of uniformity in their emergence.

In selecting seed for horticultural applications the buyer must be sure that the seed will germinate successfully (viable) and will succeed under a variety of hostile environments e.g. cold (vigour).

Viability

Viable seeds are those which retain their ability to germinate. The older the seed, the less likely it is to be alive therefore fewer seeds from old stock will germinate than seeds from new stock.

Vigour

Vigour is an estimate of the ability of the seed to germinate and develop into strong seedlings under a wide range of simulated field conditions e.g. cold and hot.

Breaking seed dormancy

The horticulturist often treats seeds to break dormancy, cause germination advancement and to improve uniform emergence. There are several techniques used for this including subjecting seeds to cycles of wetting and drying. This possibly removes germination inhibitors and improves water uptake when seed is planted.

Seed priming is a pre-sowing treatment which controls the water level within seeds to enable germination to take place. Priming has many benefits including greatly speeding up germination time, plant uniformity, enabling seeds to germinate over a wider temperature range, and it can help increase resistance to disease (e.g. carrots). Soaking seed in water or chemical primer (e.g. polyethylene glycol) can be used to start germination e.g. Mimosa.

Seed scarification is used to break dormancy of hard seed coats that will not easily allow water and air to enter (e.g. horse chestnut). These seeds will not germinate until the hard seed coat has either weathered by winter weather and soil organisms or passed through the stomach of animals. Horticulturalists force this process by rubbing the coat away with sand paper or chip with a knife or prick with a needle so that the radicle can escape.

Seed acidification involves soaking in acid between a few minutes and 24 hours e.g. scotch broom, Cytisus, magnolia.

Seed stratification: some seeds need a period of moisture and cold after harvest before they will germinate. The seed dormancy is controlled by the internal seed tissues. The treatment seems to simulate winter conditions in breaking down the seed coat. Cold stratification (moist-prechilling) treatments at 5°C for 3–5 weeks in a fridge is usually very effective e.g. *Linaria* (3 weeks), pansy (2 weeks), *Acer*, *Berberis*, *Cotoneaster*, *Pyracantha*. Some times chilling and heating is required e.g. lilies, tree paeonies.

Seed heat ripening: some seeds need a warm treatment after harvest to fully ripen them before germination. e.g. water cress.

Double dormancy: some seeds have both a tough seed coat and controls by internal seed tissues. These seeds need to be both scarified and stratified to enable germination to take place e.g. tree peonies, taxus and trillium.

The exercises in this section will explore these areas and include:

- seed structure and tissue function
- percentage germination and effect of seed age (viability test)
- embryo chemical reaction test for viability
- effect of different environments on germination (vigour test).

Table 2.1 shows some typical seed propagation characteristics.

Table 2.1 Typical seed propagation characteristics

Name	Plant temperature (°C)	Germination period (days)	Viability (years)
Vegetables			
Runner beans	18–25	10–20	2
Peas	10–20	7–15	2
Cauliflower			
Tomatoes	18–25	8–20	2
Herbs			
Basil	16–25	14–25	4–5
Watercress	6–24	12–25	4–5
Fennel	16–22	15–25	2–3
Rosemary	20–28	14–35	2–3
Annuals			
Floss flower (*Ageratum*)	22–26	10–20	2–3
Delphinium	12–16	12–20	2–3
Penstemon	18–22	15–30	2–3
Busy lizzie (*Impatiens*)	18–30	18–25	2–3
Biennials			
Hollyhock (*Alcea*)	15–22	5–12	3–4
Wallflower (*Cheiranthus*)	10–20	8–15	3–4
Foxglove (*Digitalis*)	18–22	15–25	3–4
Perennials			
Delphinium	12–18	15–25	3–4
Red-hot poker (*Kniphofia*)	18–20	15–25	3–4
Gold dust (*Alyssum*)	18–25	15–25	3–4
Poppy (*Papaver*)	10–20	15–25	2–3

Exercise 2.1

Seed structure and dissection

Background

The tissue within the seed is adapted to ensure that the embryo plant will emerge at a time to ensure maximum favourability for growth and development. For example, the tough testa seed coat prevents germination until it is broken down by the harsh winter climate. Germination is therefore prevented until the environmental conditions of spring are suitable.

Flowering plants are divided into the two subclasses: monocotyledon and dicotyledon. This refers to the number of seed leaves (called cotyledons) that the seedling establishes. Dicotyledonous plants have two cotyledons. During germination they draw on food that is stored in these seed leaves. Monocotyledonous plants have only one cotyledon and their food is stored in tissue called an endosperm. Figure 2.1 typifies the internal tissues of seeds of monocotyledon and dicotyledon plants.

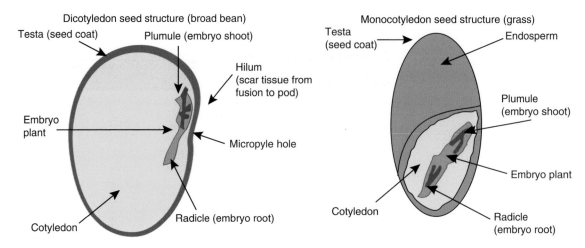

Figure 2.1 Internal seed tissue

Aim

To practise dissection and seed tissue identification.

Apparatus

White dissection tile	Scalpel
Bean seeds	Grass or wheat seeds
Seed structure illustration	Hand lens

Useful websites

http://www.cartage.org.lb/en/themes/sciences/BotanicalSciences/
PlantReproduction/PlantPropagation/SeedStructure/SeedStructure.htm

http://vegmad.org/msbp/scitech/ecophys_morph.htm

Method

1. Soak 20 seeds of each plant in water overnight to soften the tissues.
2. Bisect the bean seed and dissect the grass seed as shown in Figure 2.1.
3. Draw a labelled diagram of all the tissues found.

Results

Draw a labelled diagram of the observed seed structures.

Conclusion

1. What can you conclude about:
   (a) seed dissection
   (b) observation of tissue within the seed?
2. Name the tissue coating containing the seed.
3. State the name for the scar tissue formed where the seed was previously attached to the pod.
4. State the technical name for seed leaves.
5. What is the embryo shoot called?
6. What is the embryo root called?
7. Distinguish between seed viability and vigour.

Exercise 2.2

Percentage germination

Background

After the seed swells, having absorbed water through the micropyle through a process called osmosis (see Chapter 7), germination occurs. The first signs of germination is the emergence of the radicle (or primary root) emerging from the seed, followed by the plumule and cotyledons. The cotyledons (seed leaves) supply food to the growing plant embryo for the production of new cells. Food is stored in the cotyledons as starch. Starch is changed into soluble sugar, by enzymes as soon as the seed starts to germinate. The new plant will manufacture its own food by the process of photosynthesis as soon as it has grown tall enough to put out leaves (true leaves). Until then the cotyledons supply energy. As their food reserves are used up the cotyledons shrivel and fall off.

Figure 2.2 Epigeal germination in sunflower

Two types of germination exist: epigeal and hypogeal. In **epigeal germination** the cotyledons appear above ground. The epicotyl is the name for the emerging growing point above the cotyledons. The cotyledons are drawn up through the soil, turn green and provide a food reserve until leaves are put out and the plant starts to photosynthesize, e.g. tomatoes, cherry, courgette. Figure 2.2 shows epigeal germination in a sunflower.

In **hypogeal germination**, the cotyledons remain below the surface and the epicotyl elongates, thrusting the young shoot (plumule) above the soil level, e.g. peas, broad beans, peaches, runner beans, grasses and cereals. Figure 2.3 shows hypogeal germination in a broad bean (*Vicia faba*).

Figure 2.3 Hypogeal germination in a broad bean

Aim

To gain competence in calculating percentage germination.

Apparatus

> Oil seed rape seedlings
> Calculator
> Compost tray

Useful websites

http://theseedsite.co.uk/germinating.html

www.rhs.org.uk/Learning/Publications/pubs/garden0902/lilies.htm

Method

1. Sow the compost tray with 40 seeds of oil seed rape.
2. Allow 7 days for germination to take place.
3. Count the number of seeds that have germinated.
4. Calculate the percentage germination rate using the following formulae:

$$\text{percentage germination rate} = \frac{\text{number of successful seeds germinated}}{\text{total number of seeds sown}} \times 100$$

Results

Enter your results in the table provided.

Number of seeds sown	Number of seeds germinated	Percentage germination

Conclusions

1. Through which tissue does water enter the seed prior to germination?
2. What can you conclude about the viability of the oil seed rape seed?
3. How many cotyledons are present in the growing seedlings?
4. Explain whether oil seed rape is a monocotyledonous or dicotyledonous species.
5. State the functions of the cotyledons.

Exercise 2.3

Effect of seed age on germination

Aim

To investigate the effect of seed age on germination success.

Apparatus

    Petri dishes
    Filter paper
    Old and new lettuce seeds
    Old and new bean seeds
    Calculator

Useful websites

www.niab.com/services/laboratory-services/test-detail-pages.html

http://vegmad.org/msbp/faq/longdoseedslive.htm

Method

1. Place a disc of filter paper in a petri dish and damp it down with water.
2. Sow petri dishes A and B with 10 old and new lettuce seeds, respectively.
3. Sow petri dishes C and D with 10 old and new bean seeds, respectively.
4. Monitor and keep the filter paper moist over the next 7 days.
5. After 7 days count the total number of seeds sown and calculate the % germination rate.

Results

Enter your results in the table provided.

Petri dish	Number germinated	Total seeds sown	% germinated
A			
B			
C			
D			

Conclusions

1. Is this test for viability or vigour?
2. What can you conclude about the effect of seed age on germination?
3. How might your results influence your choice of seed storage conditions?

Exercise 2.4

Tetrazolium test

Background

Conducting germination tests on seeds can be time-consuming. As an alternative method to sowing seeds and recording resulting growth, a chemical test may be used.

The embryos of seeds contain enzymes responsible for aiding germination. These enzymes react with a colourless 1% triphenyl tetrazolium chloride solution to produce a brilliant red stain. This is referred to as the 'tetrazolium test'. Seeds that will germinate normally are stained red over the whole embryo from shoot to root. The more enzymes that are present, the greater and more widespread is the red stain, indicating germination potential. The test works equally well in dormant and non-dormant seeds.

The method is fast, and results can be obtained normally within 24 hours. The test is suitable for any seed. Most chemists and the educational suppliers listed in the back of this book will be able to supply the tetrazolium salt compound called '2,3,5-triphenyltetrazolium chloride'. Example embryo staining characteristics are given in Figure 2.4.

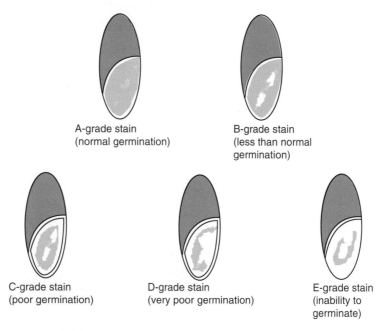

A-grade stain
(normal germination)

B-grade stain
(less than normal germination)

C-grade stain
(poor germination)

D-grade stain
(very poor germination)

E-grade stain
(inability to germinate)

Figure 2.4 Staining regions and germination potential in monocotyledon seed embryos

Aim

To test germination potential of seeds using the tetrazolium test

Apparatus

    20 wheat and pea seeds
    Tetrazolium salt
    Graph paper

Useful websites

www.defra.gov.uk/planth/pvs/guides/seedcert_stz.pdf

www.sasa.gov.uk/seed_testing/osts/adv_testing.cfm

Method

1. Prepare a 1 per cent Tetrazolium salt solution (i.e. 1 g of salt mixed with 100 ml of water) in accordance with the manufacturer's instructions.
   (a) Soak seeds overnight in water.
   (b) Soak in 1 per cent triphenyl tetrazolium chloride solution, and store in darkness for 4 hours.
   (c) Drain off tetrazolium solution, rinse with water and examine seeds within 2 hours.
2. Cut open 20 seeds of each species and observe the red stained embryos.
3. Observe the extent of the embryo staining and place each seed into a germination class.

Results

1. Categorize each examined seed into A–E grade as indicated in Figure 2.4, entering your results in the table provided.
2. Produce a bar chart summarizing your results on graph paper

Extent of embryo staining	Wheat seed Implied germination potential	Number in grade class	Percentage of total
A-grade	normal germination percentage		
B-grade	less than normal germination percentage		
C-grade	poor germination percentage		
D-grade	very poor germination percentage		
E-grade	inability to very poor germination		

Extent of embryo staining	Pea seed Implied germination potential	Number in grade class	Percentage of total
A-grade	normal germination percentage		
B-grade	less than normal germination percentage		
C-grade	poor germination percentage		
D-grade	very poor germination percentage		
E-grade	inability to very poor germination		

Conclusions

1. Is this test for viability or vigour?
2. At what concentration is Tetrazolium salt mixed in distilled water?
3. How fast can germination results be achieved using this method?
4. Calculate a total germination percentage for the sample investigated.

Germination environment

Background

All seeds are tested across a wide range of environmental conditions for their ability to germinate. This exercise investigates some of these properties.

Aim

To assess the effect of different environments on seed germination.

Apparatus

Lettuce and oil seed rape seed
Petri dishes
Filter paper
Calculator

Useful websites

www2.warwick.ac.uk/fac/sci/whri/research/seedscience/microarray/

www.kew.org/msbp/what/knowledge/germination.htm

Method

1. Place a disc of filter paper in each petri dish and damp it down with water.
2. Sow petri dishes A, B and C with 10 seeds of oil seed rape and store in 5°C (fridge), 20°C (room temperature) and 30°C (incubator or airing cupboard), respectively.
3. Sow petri dishes D, E and F with 10 seeds of lettuce and store in 5°C, 20°C and 30°C, respectively.
4. Monitor and keep the filter papers moist over the next 7 days.
5. After 7 days count the number of each seed germinated and measure the length of the plumule.

Results

Enter your results in the table provided on the next page.

Germination environment oil seed rape	Total seeds sown	Number germinated	% germinated	Average plumule length
5°C				
20°C				
30°C				
5°C				
20°C				
30°C				

Conclusions

1. Is this test for viability or vigour?
2. Explain how temperature may affect the rate of seed germination.
3. What action can a professional horticulturalists take to increase the germination rate of seeds?
4. Design a suitable experiment to further test vigour ability under different environmental conditions, stating clearly your framework.

Answers

Exercise 2.1. Seed structure and dissection

Results

Diagram of tissues similar to Figure 2.1.

Conclusions

1. (a) Can be tricky. Important to have sharp tools and well-soaked seed.
   (b) Takes practice to begin to see tissue. Rarely looks as clear as the textbook.

2. Testa

3. Hilum

4. Cotyledons

5. Plumule

6. Radicle

7. Viability is the percentage that are capable of germinating. Vigour is an ability to germinate under a range of different environmental conditions.

Exercise 2.2. Percentage germination

Germination results may vary between 80 and 100%.

Conclusions

1. Micropyle.

2. Highly viable. Normal germination should result.

3. Two

4. Dicotyledon. The leaves are net veined (see Table 1.3).

5. To fuel growth until the young seedling puts out leaves and starts to photosynthesize.

Exercise 2.3. Effect of seed age on germination

Old seeds should have significantly less success than new.

Conclusions

1. Viability.

2. Older seed is less viable.

3. Dry seed: usually 4–6 per cent moisture will survive twice as long storage as 10 per cent. Viability is also prolonged by low temperatures. Moisture-proof seed packets retain seed viability for several years.

Exercise 2.4. Tetrazolium test

Results and bar graph will vary with specific seed batch used.

Conclusions

1. Viability

2. 1 per cent

3. Within 24 hours

4. Total number germinated divided by total number sown times by 100.

Exercise 2.5. Germination environment

Conclusions

1. Vigour

2. Low temperatures slow germination down. High temperatures encourage germination. However, if temperatures are too high the seed may die.

3. Examples from background section include scarification, acidification, priming and stratification.

4. Appropriate experiment designed.

Chapter 3 Propagation

Key facts

1. Plant reproduction may be sexual or asexual.
2. Development is moving from one stage to the next in the plant's life cycle.
3. There are four plant life cycles: ephemeral, annual, biennial and perennial.
4. Vegetative propagation may occur through specialist organs including rhizomes, stolons, bulbs and corms.
5. It is easy to take cuttings from dicotyledonous plants because their stems have specialist grown tissue called cambium.

Background

The reproduction of plants is called propagation. It may be either **sexual**, reproducing from seed, or **asexual** reproduction from existing plant tissue (**vegetative propagation**). The most difficult plants to propagate are also the most expensive.

Life cycle of a plant

The stages of a plant's life cycle are:

1. Seed
2. Germination
3. Seedling
4. Young plant
5. Mature plant
6. Flowering
7. Pollination
8. Fertilization
9. Fruiting
10. Seed dispersal.

Growth occurs between each of these development stages. Growth is defined as an increase in size. Development is a stage in the life cycle of the plant. Several different life cycles exist. **Ephemerals** have very short life spans and complete several life cycles in a season. This group often include prolific weeds, e.g. Chickweed, *Matthiola, Caltha palustris* and Ragwort. **Annuals** complete one life cycle in one year, e.g. *Gypsophila* and *Sedum*. Hardy annuals are cold resistant, e.g. *Papaver commutatum*, pot marigold nasturtium, and corncockle. Usually these are grown in a greenhouse to get high-quality plants quicker than sowing them direct into the soil. Half-hardy annuals will die in cold weather and so need frost protection e.g. petunia, *Gypsophila*, busy lizzy, Salvia and pansies. **Biennials** do not flower until the second year, after which they self-seed and die, e.g. carrots, hollyhock, foxglove (see Figure 3.1). A large number of biennials are sold as autumn bedding plants e.g. wallflower.

Figure 3.1 Foxglove has a biennial life cycle

Figure 3.2 Primulas have a perennial life cycle

Perennials live for several years e.g. *Anemone*, Alstroemeria, Abutilon, Forget-me-not. Table 3.1 shows the life cycle characteristics of each group.

Table 3.1 Life cycle characteristics

Life cycle	Winter (too cold for growth)	Spring	Summer	Autumn	Examples
Ephemeral	Seed dormant	Germination, growth, flowering, dispersal, death	New generation seeds germinate. Plants flower, shed seed and die	New generation seeds germinate. Plants flower, shed seed and die	Groundsel, Shepherd's purse, Chickweed
Annual	Seed dormant	Seed germinates	Flowering	Seeds shed, plants die	Red dead-nettle, Sun flower, Burning-bush, Cornflower, Stocks
Biennial	Seed dormant	Seeds germinate	Perennating organs produced (e.g. tap root, rhizome etc.)	Perennating organs produced (e.g. tap root, rhizome etc.)	Bramble, Persian violet, Incense plant, Honesty plant, Ornamental cabbage
Second winter	perennating organs remain dormant				
Second spring		Perennating organs sprout into growth. Leaves produced			
Second summer				Flowers produced	
Second autumn				Plant dies	
Perennials	Arial parts may be dead or dormant. Roots, stems and buds are dormant	New growth	Flowering	Seed produced and dispersed. Aerial parts may die off. The whole cycle is repeated for many years	Snap dragon, New Zealand Flax, Bracken, docks, african violets, dumb-cane, asters

Sexual propagation

Sexual propagation is mainly used for plant species where the variation of characteristics (e.g. size, colour, texture) is not important. Unlike dicotyledons, monocotyledons do not have the new growth producing cells called the **cambium** layer in the stem. The cambium cells are meristem (growth) points from which new roots would emerge. Since it would therefore be very difficult to take cuttings from monocotyledons,

Figure 3.3 Violas hybridized into F1 cultivars give uniform appearance

reproduction from seed is necessary. Sexual methods are therefore essential for the propagation of **annual** plants (which live for only one season) and **biennial** plants (which are planted one year and flower the next before dying). To improve plant characteristics, for example in relation to resistance to disease, plants have to be bred by crossing with other stock to produce a superior cultivar (sexual propagation). This is called hybridization and involves raising the seed from cross-pollinated plants. New and improved plants can therefore only be created by raising seed. But once produced, their population size may be rapidly increased through taking cuttings, thus maintaining varieties and supply.

Vegetative propagation

Many plants, including trees and shrubs, are able to vegetatively propagate themselves, rather than reproduce by seed production. Other plants may require artificial methods (e.g. taking cuttings) and need great care if they are to be successful. Almost all plants sold – perennials, bulbs, corms, trees and shrubs – are vegetatively propagated. This is not only for convenience, but because most plants are hybrids which will not breed true from seed. When seeds of these plants are sown they will produce a mixture of size, shape and colour features. Vegetative propagation is used for plants where identical plants to the mother are required and variation in characteristics is not desired. These are called **clones** because they are genetically identical plants.

Plant organs and propagation

The plant is composed of four primary organs (roots, stems, leaves and flowers), all of which can be used as material for propagation. Many plant organ modifications exist to enable natural vegetative propagation. Of these, the stem is worthy of special note. The most significant aspect of stems lies in their modifications as organs associated with asexual reproduction. Such vegetative propagation results from the growth of a bud on a stem. The bud produces a completely new plant with roots, stems and leaves. All daughter plants produced are identical 'clones' of the mother plant and may also serve as food stores (e.g. tubers). These enable a quick burst of growth in the spring using stored energy (e.g. iris rhizomes).

The following structures are modified stems of great importance in vegetative propagation and some will be considered in the exercises in this chapter.

Tubers	swollen underground stems, e.g. potato, dahlia, knot grass, glory lily, caladium
Rhizomes	underground stems, e.g. iris, rhus, ginger, Convallaria
Stolens	overground stems, e.g. strawberry runners, some grasses, Ajuga
Corms	compressed stems, e.g. crocus, cyclamen, gladiola
True bulbs	very short shoots, e.g. Narcissus, hyacinths
Buds	condensed shoots ready to explode into growth, e.g. brussel sprouts and bulbils in Lilium

There are several methods used to increase plant numbers. These include cuttings, layering, division, budding and grafting. Of these, **cuttings**, **layering** and **division** will be investigated in the following exercises.

Exercise 3.1

Vegetative propagation by cuttings

Background

Taking cutting should result in the production of new plants from pieces of stem, root or leaves of older plants. The daughter plant resulting from a cutting is a **clone** of the mother. There are several methods but the primary aim is to encourage the cut tissue to produce first roots and subsequently a shoot. If new shoots grow away before the root system has developed there will be an inadequate supply of water and the plant will die. Many of the methods used therefore concentrate on ways to increase the environmental conditions under which a vigorous root system will develop e.g. applying basal root zone warming (bottom heat) and using **hormone** treatments.

Figure 3.4 These leaf cuttings are grown under polythene to slow down stem growth until the roots have formed

Example methods include hardwood cuttings (rooting increased under mist thus keeping shoots short), softwood cuttings (root in peat, sand, soil and water), leaf cuttings (e.g. begonias and saintpaulias) and root cuttings (an occasional technique e.g. horse-radish and hollyhocks). Suitable plants to take cuttings from include shrubs, conifers, many herbaceous and alpines (but not monocotyledons), some trees, some roses and climbers.

Good hygiene is essential to achieve superior results. Always use disease-free plants, clean equipment and fresh compost.

Softwood cuttings

Softwood cuttings are taken first thing in the morning in summer. Only the healthiest shoots should be selected, removing any flower buds. Figure 3.5 outlines the procedure. Some plants, e.g. hydrangea and fuchsia, prefer internodal cuttings. Suitable plants to use include geraniums, fuchsias, other tender perrenials (*Verbena, Gazania, Cosmos astrosanguinea, Argyranthemums* and *Alonsoe*), and climbers (*Plumbago, Bougainvillea, Campsis*).

Hardwood cuttings

Hardwood cuttings are taken in early autumn just after leaf fall; any earlier and too much abcissic acid is present which inhibits root formation. It is essential

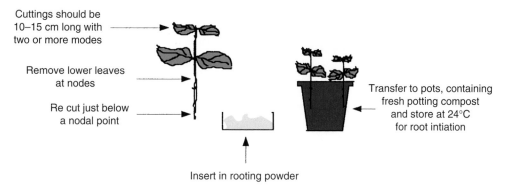

Cuttings should be 10–15 cm long with two or more modes

Remove lower leaves at nodes

Re cut just below a nodal point

Insert in rooting powder

Transfer to pots, containing fresh potting compost and store at 24°C for root intiation

Figure 3.5 Taking a softwood cutting

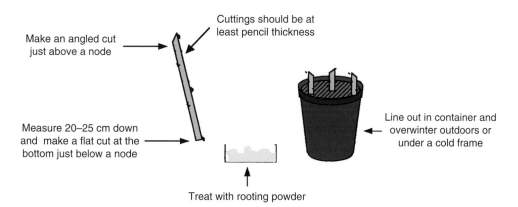

Make an angled cut just above a node

Cuttings should be at least pencil thickness

Measure 20–25 cm down and make a flat cut at the bottom just below a node

Treat with rooting powder

Line out in container and overwinter outdoors or under a cold frame

Figure 3.6 Taking a hardwood cutting

to always use good-quality, disease-free material and clean tools. Any soft whippy wood should be discarded. Figure 3.6 outlines the procedure. Suitable plants to try include, dragon-claw willow (*Salix matsudans* 'Tortuosa'), purple-leaf plum (*Prunus cerasifera* 'Atropurpurea'), *Forsythia*, *Clematis*, *Jasmine*, *Ceanothus*, and *Berberis*.

Using rooting hormones

The principal aim is to increase the percentage of cuttings that 'take' and therefore grow vigorously resulting in a healthy daughter plant.

The benefits are:

- stimulation of root initiation
- a larger percentage of cuttings form roots
- faster rooting time.

Growth regulators stimulate the rooting of easy-to-root species. Hard, or failure-to-root species, normally show no response to hormone treatment.

Cuttings and rooting hormones

The hormone called auxin (IBA, Indole-butyric acid), is the most common rooting stimulant. This is because it is very persistent (long-lasting), due to weak auxin activity, and effective as a root promoter. Because IBA translocates (moves around the plant) poorly, it is retained near the site of application. Hormones that readily translocate may cause undesirable growth effects in the propagated plant.

NAA (naphthylacetic acid) is also very proficient at root promotion. However, it is more toxic than IBA, and excessive concentrations are likely to damage or kill the plant.

The naturally occurring hormone is called IAA (indole acetic acid). Both IBA and NAA are more effective at promoting roots than IAA, which is very unstable in plants. Decomposition of IAA occurs rapidly, unlike man-made compounds.

Growth regulators may alter the type of roots formed as well as the quantity grown. IBA produces a strong fibrous root system whereas NAA often produces a bushy but stunted root system.

Root-promoting compounds work better when used in combination with others. Fifty per cent IBA plus 50 per cent NAA results in a larger percentage of cuttings rooting than either material used alone.

Method of application

There are three methods that work particularly well for stem cuttings.

Quick dip

The basal end of the cutting is dipped into a concentrated solution (500 to 10,000 parts per million, ppm) of the chemical dissolved in alcohol. Cuttings are dipped for varying lengths of time ranging from 5 minutes to 6 hours, depending on:

(a) the species
(b) type of cutting
(c) age of tissue
(d) concentration of solution.

Solutions of auxin can be stored for fairly long periods of time provided they are in a closed container, out of direct light and chilled.

Prolonged dip

A concentrated stock solution is prepared in alcohol. This is then diluted in water at the time of use to give strengths of 20–200 ppm. Impurities in the water rapidly help to decompose the auxin and render it useless. The cuttings are soaked in this solution for up to 24 hours. Uptake of auxin tends to be more erratic and depends on the environmental conditions. It is therefore inconsistent.

Powder dip

The growth hormone is contained in an inert powder (e.g. clay or talc), at concentrations of 200–1000 ppm (softwood cuttings) or 1000–5000 ppm (hardwood cuttings). The cutting is dipped in the rooting powder and then planted. The auxin is then absorbed from the powder.

Aim

To assess the effectiveness of different hormones on the percentage 'take' of cuttings.

Apparatus

Geranium stock plants
Seed tray
IBA solution (100 ppm)
Scalpel
NAA solution (100 ppm)

Compost
IAA solution (100 ppm)
Small beakers
50 per cent IBA and NAA solution (100 ppm)
Plant labels

Hormone rooting powder e.g. Murphy Hormone rooting powder (captan + 1-naphthylacetic acid)

Useful websites

www.rooting-hormones.com/

www.rhs.org.uk/publications/pubs/garden0405/propagation.asp

Method

1. Prepare the hormone solutions to the required concentrations, by dilution with distilled water.
2. Select three of the five hormone treatments for study.
3. Prepare 3 stem cutting of the Geranium plant. You may collect your own additional cuttings (e.g. woody species) for investigation if you wish.
4. Label each cutting with your initials, date and hormone treatment under investigation.
5. Using the 'Quick dip' technique previously described, treat your cutting and place into a compost seed tray applicable for that treatment.
6. Record weekly the effectiveness of the treatments in the table provided.
7. On your next visit to the garden centre, record the type of hormones used in amateur rooting compounds.

Results

Enter your results in the table provided.

Cutting treatment at 100 ppm	Rooting Response			Total cuttings treated	% Successful take
	Week 1	Week 2	Week 3		
IBA					
NAA					
IAA					
50% IBA and NAA					
Hormone rooting powder					

Conclusions

1. Why is an inert powder used?
2. Why do hardwood cuttings require a more concentrated hormone application?
3. Why is distilled water used?
4. Why is the daughter plant identical to the parent?
5. What is meant by breeding true?
6. What benefits are there from hybridization?
7. State the naturally occurring plant hormone.
8. What effect can an over-application of NAA have on cuttings?
9. What advantages do man-made compounds have over the naturally occurring plant hormones?

10. Calculate the percentage of cuttings that formed roots.
11. State what the initials 'ppm' mean.
12. What other techniques can you state that increase the effectiveness of cuttings to root?

Exercise 3.2

Propagation and layering (stem rhizomes and stolons)

Background

Plant stems are often suitable propagation material to increase plant stock. Strawberry plants, for example, are often supplied in this form as 'runners'. In addition many weeds owe their success to an ability to evade some herbicides by reproducing this way (see also Exercise 3.3).

Rhizomes are underground horizontal stems growing from a lateral bud near the stem base. At each node there is a bud capable of sending out roots and shoots (e.g. lilly-of-the-valley, sand sedge, couch grass, canna lily, ginger, zandeschia, solomon's seal, rhus and bulrush).

Figure 3.7 shows the anatomy of an iris rhizome.

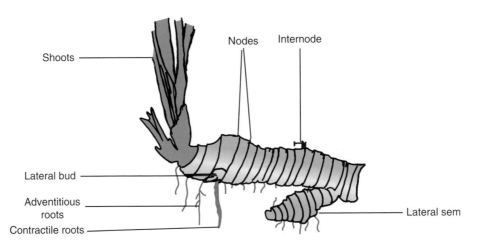

Figure 3.7 The anatomy of an iris rhizome

Stolons are overground stems. The stems bend over and form new plants at their tips (offsets). They have very long internodes with small scale leaves at the nodes. There is a bud at each node which can produce both roots and shoots. Eventually the stolon dies, leaving an independent daughter plant (e.g. brambles, ground ivy, creeping buttercup, houseleek, strawberry and blackberry).

Figure 3.8 shows the anatomy of a strawberry stolon.

Layering

Stolon-forming plants naturally layer themselves (e.g. rhododendron and honeysuckle). New plants are formed when stems make contact with the surrounding soil where a new root system develops. This technique is suitable for most woody plants and shrubs. Stems are bent over and staked to the surface of the ground, thus encouraging the plant to root and shoot while still attached to the mother plant. The stolon can then be severed and the

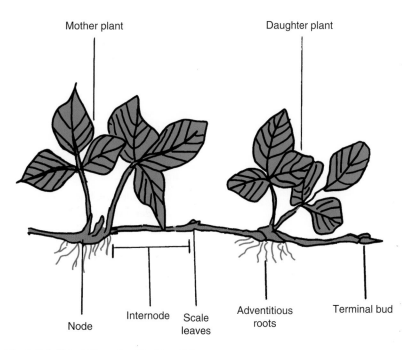

Figure 3.8 The anatomy of a strawberry stolon

new plant grown in situ or potted-on (e.g. climbers, carnations, strawberries, jasmine).

Aim

To distinguish between rhizome and stolon stems as agents of vegetative propagation.

Apparatus

   Iris and couch rhizomes
   Strawberry runners

Useful websites

www.greenfingers.com/articledisplay.asp?id=237

www.botgard.ucla.edu/html/botanytextbooks/generalbotany/typesofshoots/stolon/index.html

Method

Observe the specimens

Results

Draw a labelled diagram of the couch rhizome and strawberry stolon including the following terms if relevant:node, internodes, shoots, stolon, rhizome, mother plant, adventitious roots.

Conclusions

1. From where did the 'cloned' daughter plant form?
2. Define a rhizome.
3. Define a stolon.
4. Plants that have rhizomes are often resistant to flame weed killers. Explain why this is so.

5. **On BBC Radio Four's** *Gardener's Question Time* **Mrs Davies asked why pulling up couch grass was not effective in eliminating the weed from her garden. What would be your answer if you were on the panel?**

Exercise 3.3

Propagation and division (stem corms and bulbs)

Background

Corms and bulbs are further examples of modified stems and have a particular ability to perennate (survive over winter), thus giving good growth early in the next season.

Corms are underground vertically compressed stems which bear buds in the axils of the scale-like remains of the previous season's leaves (e.g. crocus, gladiola, cyclamen, anemones, freesia and cuckoo pint). These are similar to bulbs but it is the stem that swells rather than the leaf bases.

Figure 3.9 shows the anatomy of a crocus corm.

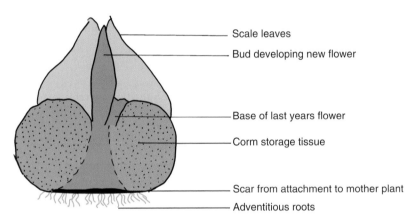

Figure 3.9 The anatomy of a crocus corm

True bulbs are underground swollen fleshly leaved storage bases encircling a very short shoot (e.g. narcissus, lily, tulip, nerine, hyacinth, onions, snow drop and bluebell). A bulb contains the whole plant including the flower head in immature form. In spring the rapidly growing terminal bud uses the stored food in the swollen leaves to produce a flowering stem and leaves. Energy manufactured from photosynthesis is stored in the leaf base, which then swells forming a new bulb to perennate the next winter.

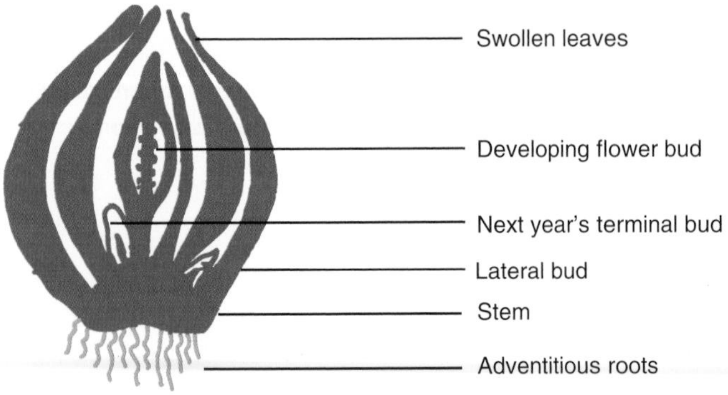

Figure 3.10 shows a tulip bulb anatomy.

Figure 3.11 shows a daffodil bulb dissected; note the developing flower bud and clearly seen axillary buds.

Figure 3.10 Tulip bulb anatomy

Figure 3.11 Daffodil bulb dissected

Division

This involves separating the root stocks to produce two new plants. This method is only possible with bulbs, herbaceous perennials and plants producing rhizomes or stolons. The growth medium should be thoroughly wetted up first. Remove the mother plant and shake off excess soil or compost. Then divide the root ball by either hand, two back-to-back forks or a sharp knife depending on the toughness of the root ball.

Best results are achieved when the plant is dormant during the winter months. Suitable materials include herbaceous and alpine plants. This technique is particularly important commercially for the propagation of raspberries and strawberries.

Division and plant organs

Many monocotyledons plants, and others which have fibrous root systems, can generate new plant material simply by dividing the plant at the root clump. Plants that reproduce by rhizomes and stolons can also be divided provided the stem contains a bud (e.g. *Stachys*, *Aster* and some ornamental grasses). The divided material is then simply potted-up, often taking the opportunity to cut back some foliage at the same time.

Suitable plants include herbaceous perennials (Aster, Hosta, Stachys), perennial herbs (chives and mint), alpine and rock plants (Aubretia, Dianthus, Phlox, Sedum), house plants (Acorus, Aspidistra, Caladium, Chlorophytum, Cyperus, Saintpaulia, Scirpus and stag's horn ferns), and succulents (Aloe, Agave, Echeveria).

For best results house plants should be divided in spring before new growth starts. Early flowering species should be divided in mid summer immediately after flowering. Late flowering species should be divided between October and March (see also Exercise 4.3).

Aim

Investigation of bulbs and corms as tissue for propagation by division.

Apparatus

    Crocus corms
    Daffodil bulbs
    Dissection kits

Useful websites

http://theseedsite.co.uk/bulbs.html

Method

1. Bisect the specimens.
2. Observe the corms' axillary buds and swollen stems.
3. Observe the bulbs' folded scales (leaves), flower and stem.

Results

Draw a labelled diagram of your observations, showing clearly how corms differ from bulbs to include the following terms if relevant: flower bud, swollen leaves, future leaves, last year's leaf base, axillary bud, stem, adventitious roots.

Conclusions

1. Corms, tubers and rhizomes are often wrongly described as bulbs. Explain the differences.
2. Explain the advantages and disadvantages of corms and bulbs to plants in perennation.
3. What features make bulbs suitable as material for vegetative propagation?
4. Why are corms not suitable as material for vegetative propagation?

Exercise 3.4

Propagation and division (bulbs)

Aim

To utilize previously dissected bulbs from Exercise 3.3; to produce vegetatively propagated offspring using the 'chipping' method.

Apparatus

Dissected bulbs (e.g. Narcissus)
Scalpel
9:1:1: vermiculite, peat, water media
Incubator 23°C
0.2 per cent Benlate fungicide solution

Useful websites

www.intergardening.co.uk/practical/propagation/dividing-bulbs/chipping.html

Method

1. Trim of the nose and stem of the dissected bulb.
2. Cut the bulb longitudinally into quarters.
3. Cut off slices from each quarter about 4 mm thick on the outside tapering into the centre.
4. Place slices (chips) into 0.2 per cent Benlate fungicide solution for 30 minutes.
5. Place chip in media tray, recording how many are inserted.
6. Incubate for 2 months at 23°C.
7. After 2 months small bulbs will have developed ready for planting out.

Results

1. Record the total number of incubated chips produced.
2. After 2 months record the number of successful bulbs.
3. Calculate the percentage success rate.

Conclusions

1. Explain the variation in results between species.
2. Why do you think some chips were unsuccessful?
3. How might the percentage success rate be improved?
4. List other methods of vegetative propagation with which you are familiar.

Answers

Exercise 3.1. Vegetative propagation by cuttings

Conclusions

1. So that conditions are clean and disease free. The powder also helps to carry and evenly distribute the hormone.

2. Cells contain more lignin and are less able to absorb the hormone compared to the young growth of softwood cuttings.

3. It is the purest form of water available.

4. Because it has been vegetatively propagated and is a clone of the mother plant.

5. It leads to improved characteristics such as resistance to pest and disease.

6. Plants raised from seed having all the qualities of the species grown. Often e.g. *Pyracantha*, cross-fertilization will occur and the seed will contain a hybrid plant.

7. Auxin occuring as idole acetic acid (IBA).

8. Damage or kill the plant.

9. Decomposition occurs slowly and applications are therefore more effective.

10. Calcultation using formulae: Number of successful cuttings/total number planted $\times$ by 100

11. Parts per million.

12. Use bottom heat. Mist propagation, fleece coverings, etc.

Exercise 3.2. Propagation and layering

Diagrams as Figures 3.5 and 3.6.

Conclusions

1. Nodal points.

2. Underground horizontally spreading stem.

3. Overground horizontally spreading stem.

4. Only the arial part is killed by the flame gun. The underground parts contain stored energy to continue growth.

5. Couch grass reproduced by rhizomes. Although Mrs Davies might have removed what aerial parts she could see, there will still be a considerable amount of underground stem remaining.

Exercise 3.3. Propagation and division (stem corms and bulbs, refer to Figures 3.5 and 3.6)

Conclusions

1. Corms are vertically compressed stems. Tubers are swollen underground stems used as food storage organs. Rhizomes are underground horizontally spreading stems. Bulbs are very short shoots surrounded by fleshy leaves. The fleshy leaves contain energy reserves which will fuel the bulbs' rapid flower growth in spring.

2. Corms contain carbohydrate storage areas ready to fuel growth in the spring until the new plant produces its own leaves and starts to photosynthesize. A bulb's stored energy is contained in the swollen fleshy leaves which fuel early growth, enabling the stem to grow above the competition and begin photosynthesizing.

3. The bulb can be dissected as in Exercise 3.4, and each part is cable of growing into a new plant.

4. New growth does not arise from dissection although some plants, e.g. crocus, do produce tiny separate cormlets at the end of the season.

Exercise 3.4. Propagation and division (bulbs)

Results will vary with hygiene conditions and the experience of the operator.

Conclusions

1. Tulips seem to give better results than daffodils. This may be because the bulb is more compact than the daffodil.

2. Hygiene may have been less than satisfactory introducing disease. It may also be that too much genetic material was removed from the chip.

3. Be scrupulously clean, use fresh compost and vermiculite.

4. Layering, division and cuttings.

Chapter 4   Weed – biology and control

Key facts

1. A weed is a plant growing in the wrong place.
2. It is important to be able to identify weeds at both the seedling and adult stage.
3. Many weeds are able to regenerate from chopped up fragments of their roots and stems.
4. Roots are either fibrous or tap roots.
5. Techniques of cultural weed control are essential in organic and sustainable horticulture.
6. Chemical weed controls mostly use hormones to pervert plant growth leading to weed death.

Background

A weed is any undesirable plant hindering or suppressing the growth of more valued plants through competition for water, nutrients or light and may additionally harbour insects carrying viruses or disease. Figure 4.1 shows a daisy unwanted in a lawn, yet cultivated varieties of daisy with red, pink and yellow flowers are increasingly popular as bedding plants.

Figure 4.1 An unwanted daisy in a lawn

Weeds tend to invade every area of horticulture from sports turf, amenity plantings, nursery stock to glasshouse soils. The effects can be devastating, especially if they bring other pests and disease with them. They have aspects to their biology that make them successful, such as fast germination rates or mass of seed produced. In addition, control methods may vary at different growth stages. Annual weeds in particular are susceptible to contact herbicide control at the seedling stage, before the cotyledons have dropped off, but not at the adult stage. Weed identification can be a difficult skill to master and for this reason it is worthwhile starting with weeds in flower as this aids speedy identification. Weeds are normally be divided into annual and perennial weeds although examples of ephemerals and biennials can also be found (see also Table 3.1).

Ephemeral weeds

An ephemeral is a plant that can complete several life cycles in a year. For example, chickweed completes its life cycle in 8 weeks and will produce many generations in a year. Other examples include shepherd's purse and wavy bittercress.

Figure 4.2 Groundsel will shed 30,000 seeds

Annual weeds

These complete their life cycle, from germination to death, in one season. They are quick growing throughout the year and produce abundant seeds which can remain dormant in the soil for several years. Viability varies greatly from as little as three years for groundsel to thirty years for speedwell seeds in the soil. A groundsel weed will shed around 30,000 seeds. Hence the saying, 'One year's seeding gives seven years weeding.' Generally, ephemeral weeds are grouped together with annuals. Germination is triggered by warm temperatures, moisture and light. Examples include speedwell, oxalis, fat hen, groundsel, sowthistle, scentless mayweed, sun spurge and red dead nettle.

Biennial weeds

Bienniel weeds do not flower and set seed until their second year, after which they die. In the first year the weed grows vegetatively, storing energy reserves which are reused to flower in year two. Examples include bramble, caper spurge, evening primrose, ragwort, hogweed, spear thistle and burdock.

Perennial weeds

These plants normally flower annually and live for more than two years. They often die back during the winter and regrow when the temperature rises in spring. They are characterized by tap roots and rhizomes that regrow if chopped into small pieces such as when rotavating. For this reason they are difficult to remove from soil. Examples include bindweed, creeping thistle, couch, Japanese knotweed, yarrow, horsetails and broad-leaved dock.

Figure 4.3 Yarrow easily regrows from rotavated pieces

Exercise 4.1

Weed collecting and identification

Aim

To recognize and collect some annual and perennial weeds in the adult and seedling stages, from different horticultural situations.

Apparatus

    Hand lens
    Microscopes
    Trowels
    Sample bags

Wildflower handbook, e.g. Fitter R.S.R. (1985) *Wild Flowers of Britain and Northern Europe* (Collins); Press B., (1993) *Field guide to the wild flowers of Britain & Europe* (New Holland)

Useful websites

www.british-wild-flowers.co.uk

www.botanicalkeys.co.uk

Method

1. Using the wildflower handbooks and Figure 4.4 as a guide, collect three different annual and three different perennial weeds from one, or more, of the following horticultural situations:
    (a) in a lawn
    (b) in a flower border/bed or in a growing field crop
    (c) below established trees/shrubs
    (d) in a hard landscaped area e.g. paving or gravelled area
    (e) in a glasshouse soil.

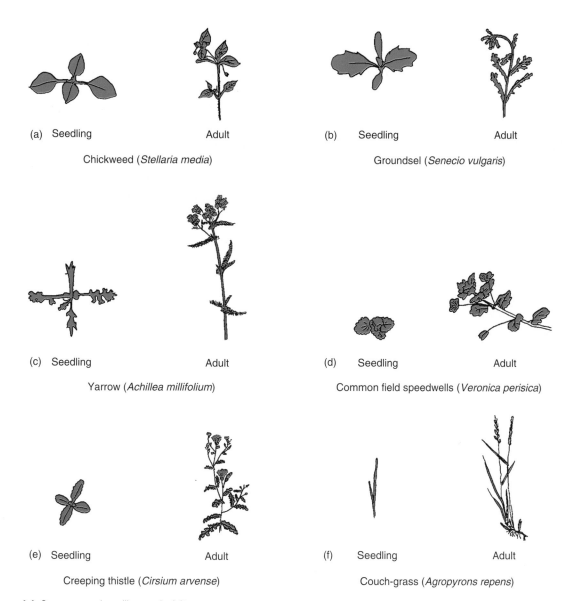

(a) Seedling   Adult
        Chickweed (*Stellaria media*)

(b) Seedling   Adult
        Groundsel (*Senecio vulgaris*)

(c) Seedling   Adult
        Yarrow (*Achillea millifolium*)

(d) Seedling   Adult
        Common field speedwells (*Veronica perisica*)

(e) Seedling   Adult
        Creeping thistle (*Cirsium arvense*)

(f) Seedling   Adult
        Couch-grass (*Agropyrons repens*)

Figure 4.4 Common weed seedlings and adults

The following weeds are suggested targets for collection (see Figure 4.4):

Annuals	Perennials
chickweed (*Stellaria media*)	creeping thistle (*Cirsum arvense*)
Groundsel (*Senecio vulgaris*)	couch grass (*Agropyron repens*)
speedwells (*Veronica* spp.)	yarrow (*Achillea millifolium*)

2. Also try collecting seedling specimens, with their cotyledons present, to see at which stage they could be controlled.
3. Use the trowel with care to dig up as much of the plant as possible including the roots.
4. Place in sample bag for later observations.
5. Observe the specimen under the microscope or using a hand lens note any significant features.
6. Using the wildflower handbook, or useful website, identify and name each weed and state the characteristics of the weed that make it successful (e.g. annual, habit of plant, acid soil lover).

Figure 4.5 Speedwell invading a lawn

7. Record the location and site features of where the weed was found (e.g. soil type, aspects of location and how this differs from site to site).
8. Save the specimens for use with subsequent exercises.

Results

Enter your results in the table provided.

Horticultural situation	Name of weed	Botanical characteristics that make the weed successful	Date found	Location found	Details of site
Lawn	1.				
	2.				
	3.				
Flower border/field crop	1.				
	2.				
	3.				
Established trees or shrubs	1.				
	2.				
	3.				
Had landscaped area	1.				
	2.				
	3.				
Glasshouse soil	1.				
	2.				
	3.				

Conclusions

1. State what is meant by the term 'weed'.
2. List four features that make annual weeds successful in cultivated areas.
3. Give an example of an annual and perennial weed.
4. List four harmful effects of weeds.
5. Were there any associations between the location of the weeds that you found and the surrounding community/landscape? If so, what?

Exercise 4.2

Pressing and mounting weed specimens

Background

Once competence has been gained at weed recognition it will be useful to keep a record to aid future quick reference and identification. A herbarium of pressed plants can be produced. One of the largest collections in the world is at Kew Gardens, which houses over seven million specimens. The Royal Botanic Gardens in Edinburgh has over three million pressed plants. A herbarium record can be achieved by pressing the plant material in a plant press or under a flat heavy object, between sheets of paper to absorb the sap. The press flattens and dries plant material. When dry they can be mounted on white card and secured with a little glue. They may then be protected by covering with 'transpaseal' or other form of sticky-back plastic. However, the pressing stage is the most critical and if not properly done, mould and other fungal growths will appear under the transpaseal, concealing any chance of later identifying the specimen. Thick and succulent materials are particularly vulnerable, but this can be easily avoided by changing the paper daily at first, then less often once most of the sap has been removed.

Aim

To produce a collection of pressed weeds for future quick reference and identification

Apparatus

Absorbent paper plant press weeds specimens (collected from Exercise 4.1)

Useful websites

www.kew.org www.rbge.org.uk

Method

Follow the instructions in Figure 4.6.

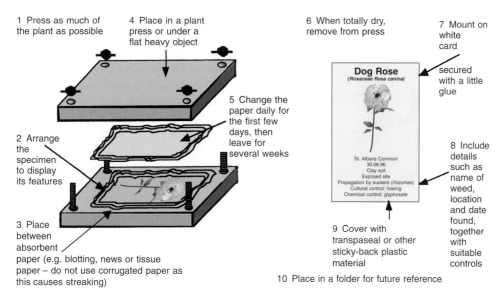

1 Press as much of the plant as possible

4 Place in a plant press or under a flat heavy object

6 When totally dry, remove from press

7 Mount on white card

Dog Rose
(Rosaceae Rosa canina)

secured with a little glue

5 Change the paper daily for the first few days, then leave for several weeks

2 Arrange the specimen to display its features

St. Albans Common
30.06.96
Clay soil
Exposed site
Propagation by suckers (rhizomes)
Cultural control: hoeing
Chemical control: glyphosate

8 Include details such as name of weed, location and date found, together with suitable controls

3 Place between absorbent paper (e.g. blotting, news or tissue paper – do not use corrugated paper as this causes streaking)

9 Cover with transpaseal or other sticky-back plastic material

10 Place in a folder for future reference

Figure 4.6 Pressing and mounting weed speciments

Results

Depending on the species drying may take several weeks to several months

Conclusions

1. How should the material be arranged on the press?
2. What causes the build up of mould on some pressed specimens?
3. Explain what difficulties pressing succulent plants presents.

Exercise 4.3

Root structure

Background

The root systems of both weeds and cultivated plants play an important role in vegetatively propagating new plants, particularly following cultivation when the root may be chopped into several pieces, each capable of producing a new plant. This exercise is designed to gain familiarization with the structure of plant roots, not only for weeds but also for general horticultural practices. This knowledge is developed in subsequent exercises. Root hairs grow between the soil particles and take up water by osmosis, and minerals by diffusion. The surface area of roots of rye has been estimated as 200 m2. The additional area provided by root hairs was 400 m^2.

The roots are the primary organ for the uptake of water into the plant, but they have other functions such as:

(a) to anchor plant to the ground
(b) to absorb water and dissolved mineral salts from the soil
(c) to act as a pipeline between the soil and the stem
(d) modified, as organs of food storage and vegetative propagation.

Most roots are either tap roots (mainly dicotyledons) or fibrous roots (mainly monocotyledons).

Tap root

These have a strong main root descending vertically with little or no lateral growth (see Figure 4.7). These roots are often strengthened by lignin and become very hard to remove and inflexible. Conifer tap roots can penetrate chalk 20–30 m deep, providing good support and resistance to drought.

Some tap roots when chopped up produce new plants (e.g. chrysanthemum and docks) and therefore make weed control very difficult. Some trees and herbaceous perennials are vegetatively propagated from root cuttings (e.g. many perennials and biennials, dandelion, dock, parsnip and carrot) (see also Propagation exercises on Division).

Fibrous root system

This is a root which branches in all directions in a mass of fine matted roots rather than thick, fleshy ones (see Figure 4.7). These explore the top soil much more effectively than tap roots. However, as they are shallow rooting, they suffer in dry weather and need irrigation. This can be difficult since water passes through the soil profile quickly and passes beyond the reach of the roots. These plants are very susceptible to drought and require accurate watering if plant loss is to be avoided (e.g. most annuals, groundsel, grass, ferns, and house plants in general).

Adventitious roots

These occur in unusual or unexpected locations, such as the negatively phototropic roots on the side of juvenile aerial stems of ivy. New roots which form from stem cuttings are also called adventitious roots. Iris rhizomes shows adventitious roots at their nodes.

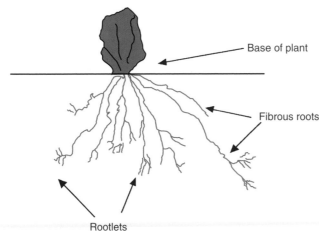

Figure 4.7 Root structures

Figure 4.8 Fibrous roots of turf cocoon the soil with a web of root hairs

Rooting depth

In summer conditions, when the soil is drying out, water may not move by capillary action fast enough towards the root system to keep pace with transpiration loss. Roots must continually grow into new damp areas of soil to maintain the water supply. Soil compaction and pans could prevent this. Roots consequently have a large surface area to improve water absorption, but are disadvantaged by being susceptible to diseases (e.g. club root).

Container-grown plant roots

Woody plants grown in containers need regular potting-on into larger containers so that roots have space to grow otherwise they become pot-bound – roots tend to spiral round the pot and plants may die, even up to twelve years after subsequent planting out.

Aim

To investigate how different root structures influence the success and management of plants.

Apparatus

Tap root from broad leaved dock
Fibrous grass roots
Adventitious roots of ivy
Binocular microscope
Annual weeds: chickweed (*Stellaria media*), groundsel (*Senecio vulgaris*), speedwells (*Veronica* spp.)
Perennial weeds: creeping thistle (*Cirsium arvense),* couch grass (*Agropyron repens*), yarrow (*Achillea millifolium*)

Useful websites

http://facweb.furman.edu/~lthompson/bgy34/plantanatomy/plant_root.htm

www.botanical-online.com/raizangles.htm

Method

1. Observe the specimen tap and fibrous roots under the microscope.
2. Observe fine root hairs under the binocular microscope.
3. Observe and classify the annual and perennial weed's roots into 'tap' or 'fibrous' classes.

Results

Enter your results in the table provided.

Life cycle type	Weed species	Root structure
Annual weeds	chickweed (*Stellaria media*)	
	groundsel (*Senecio vulgaris*)	
	speedwells (*Veronica* spp.)	
Perennial weeds	creeping thistle (*Cirsium arvense*)	
	couch grass (*Agropyron repens*)	
	yarrow (*Achillea millifolium*)	

2. Root hairs allow water to enter the root by osmosis (see Chapter 7). Estimate the length of root hairs on a 1 cm length of your specimens.
3. Draw a labelled diagram of the parts of tap and fibrous systems, including, main root, lateral root, root hairs, where appropriate.

Conclusions

1. Which root system would be more susceptible to drought conditions and why?
2. Which root system would be least susceptible to draught conditions and why?
3. Explain what precautions should be taken when growing woody plants in containers.
4. How can the roots system help you to understand how plant species become successfully adapted to their environment?
5. How would you tell which root system was the most lignified?
6. How can monocotyledons and dicotyledon species mostly be separated by their root systems?
7. List four features that make perennial weeds successful.

Exercise 4.4

Cultural weed control

Background

Generally, weed control methods are either 'cultural' or 'chemical'. Chemical control will be considered in Exercise 4.5. Cultural control refers to cultivation techniques of 'good practice': tending carefully the growing plants, and removing competitors, without using chemicals. Cultural control is essential in organic and sustainable horticulture because no man-made chemicals are introduced to the soil ecosystem. It includes the use of mechanical cultivators and rotavators. There are three main cultural control methods.

Figure 4.9 Removing moss from a lawn

Hoeing

Hoeing simply involves bringing the weed to the surface where the action of sun and the absence of water kill the plant. For this reason it is more effective in hot, dry weather. This is particularly suited for annual weeds, provided the roots are fully exposed. Hoeing and rotavating always brings more weeds to the surface through disturbing the soil. Scarification is a similar technique used to remove moss from of lawns using either a hard rake or mechanical scarifier.

Rotavating

In addition to turning the soil this practice chops weeds up into small pieces. However, perennials, are more resilient, largely because of their ability to regenerate from root cuttings, and for effective control, this practice must be continued for the whole season until the action of rotavation has exhausted the plant's food supply. This feature will be investigated in this exercise.

Flame weedkillers

This treatment has tended to be used for glasshouse crops were some degree of soil sterilization is also required. Small ride-on machines, with their own gas supply, are now available. Steam sterilization is also a common glasshouse practice. However, several portable flame guns are now available from garden centres and are becoming increasingly popular.

Other techniques include: hand weeding with a small fork to dig out roots; improving soil structure, especially drainage (e.g. spiking and aeration works); creating a stale seed bed by irrigating to encourage annual weeds to germinate which are then killed before replanting; purchasing certified clean seed that has been screened against weeds seeds; raking, so that stems are brought upwards to meet blades of cutter, or presented to hoe (e.g. to remove clover *Trifolium repens*); scarifying to removing dead organic matter preventing build up of a microclimate that favours certain weeds (e.g. moulds, and annual meadow grass); reducing shade (e.g. to discourage lesser celandine *Ranunculus ficaria* from lawns); and mulching or barriers to prevent light reaching weeds. For example ground elder can be effectively controlled by smothering with black polythene.

Figure 4.10 Hand weeding dandelions using a fork

Figure 4.11 Grass clippings recycled as a mulch barrier preventing weed growth in beds and borders

Aim

To investigate the effectiveness of different rotavation techniques to prevent the regrowth of annual and perennial weeds.

Apparatus

Soil-filled seed trays ($\times$7)
Secateurs
Annual weeds: chickweed (*Stellaria media*), groundsel (*Senecio vulgaris*), speedwells (*Veronica* spp.)
Perennial weeds: creeping thistle (*Cirsium arvense*), couch grass (*Agropyron repens*), yarrow (*Achillea millifolium*), broad-leaved dock (*Rumex obtusifolius*)

Method

1. Collect examples of the annual and perennial weeds listed above.
2. Cut the roots of each plant into lengths of 1 cm, 2 cm, 5 cm, 7 cm and 10 cm lengths.
3. Record the number of cuttings made in each size band.
4. Plant cuttings in a labelled seed tray.
5. Leave for 7 days to regrow.
6. After 7 days record the number of cuttings in each size band that have successfully regrown.

Results

Enter your results in the table provided in the next page.

	Type of weed	Cutting size-band	Number of cuttings planted	Number regrown after 7 days	Percentage regrowth	Percentage not regrown
Annuals	chickweed (*Stellaria media*)	1 cm				
		2 cm				
		5 cm				
		7 cm				
		10 cm				
	groundsel (*Senecio vulgaris*)	1 cm				
		2 cm				
		5 cm				
		7 cm				
		10 cm				
	speedwells (*Veronica* spp.)	1 cm				
		2 cm				
		5 cm				
		7 cm				
		10 cm				
Perennials	couch grass (*Agropyron repens*)	1 cm				
		2 cm				
		5 cm				
		7 cm				
		10 cm				
	creeping thistle (*Cirsium arvense*)	1 cm				
		2 cm				
		5 cm				
		7 cm				
		10 cm				
	yarrow (*Achillea millifolium*)	1 cm				
		2 cm				
		5 cm				
		7 cm				
		10 cm				
	broad-leaved dock (*Rumex obtusifolius*)	1 cm				
		2 cm				
		5 cm				
		7 cm				
		10 cm				

Conclusions

1. Why are cultivation techniques unsuitable for eliminating dock weeds?
2. State the rotavated cutting size necessary to prevent regrowth for each of the investigated weeds.
3. How effective was rotavation at eliminating the growth of annual weeds?
4. How effective was rotavation at eliminating the growth of perennial weeds?

5. Describe how the grower can affect the growth of weeds by cultivation practices.
6. State four methods of controlling weeds without using herbicides.

Chemical weed control

Chemical control

Do not use weedkillers indiscriminately. Several are toxic to beneficial insects and can do ecological damage. Herbicide use is now covered by the **Food and Environmental Protection Act** (FEPA) of 1985, and requires operators to be certified as competent users. This means that while a gardener may visit the garden centre and buy weedkillers for their own use, if a professional buys the same weedkillers to apply to a customer's garden then they must have a certificate. Resistance may build up if chemicals are persistently used and some degree of crop rotation should be practised to reduce this risk. *The UK Pesticide Guide* (Whitehead, 2008), is an essential reference source to identify suitable herbicides for use in horticulture and is reviewed annually. The use of herbicides has greatly reduced labour costs and also reduced the wide row spacing previously necessary between some crops (e.g. carrots).

Chemical weed control normally involves the use of some form of modified plant hormone as a growth regulator. For example, the use of 2,4D (a translocated herbicide) is a form of synthetic auxin. When sprayed on dicotyledonous weeds (e.g. thistles) it causes abnormal growth. The inter nodes become severely elongated, the phloem becomes blocked and the meristem points produce mutated cells. The root system becomes corrupted and death follows swiftly thereafter.

They may be classified as follows:

- non-selective – kill all plants whether weeds or crops. Being non-selective it is essential to avoid spray drift e.g. glyphosate.
- selective – kill weeds but do not harm crops. This is called '**differential toxicity**'. Different groups of plants have different responses to the same concentration of weedkiller e.g. dicotyledons are more sensitive than monocotyledons to 2,4D. Thus when sprayed on to a grassed lawn (monocotyledons), thistle weeds (dicotyledons) die off. Similarly with Propyzamide herbicide; some examples of the concentration (in parts per million) required to cause plant death are given below to demonstrate differential toxicity are: crop – carrot (0.8 ppm), cabbage (1.0 ppm), lettuce (78.0 ppm); weeds – knot grass (0.08 ppm), fat hen (0.2 ppm), groundsel (78.0 ppm).
- contact – kill only those parts that they touch.
- translocated – absorbed and travel through the phloem to kill the meristem points in both roots and shoots. This method is essential to kill the root system of large established weeds e.g. glyphosate.
- non-residual – act immediately after application e.g. paraquat.
- residual – remain active in the soil for long periods of time. This type of weedkiller is popular to protect patios and drives from weed invasion, e.g. sodium chlorate and dichlobenil.
- foliage acting – absorbed through the leaves.
- soil acting – absorbed through the roots.

Some herbicides have different characteristics with varying plants. Some herbicides, for example, are absorbed through both leaves and roots; others are selective in one situation and non-selective in another. The roots and shoots of the same plant will respond differently to the same concentration of weedkillers. In fact many herbicides contain a mixture of the various different chemicals to overcome this.

Aim

To select suitable herbicides to control annual and perennial weeds at different growth stages, in a range of horticultural situations.

Apparatus

The UK Pesticide Guide ('the green book'; Whithead, 2008)

Useful website

www.ukpesticideguide.co.uk

Method

Use the current edition of *The UK Pesticide Guide* to find suitable herbicides details to control the following annual and perennial weeds, at both the seedling and adult stage.

Results

Enter your results in the table provided.

Horticultural situation and name of weed	Growth stage	Active ingredient	Product name	Manufacturing company	Precautions to be taken when using this product
Lawn: yarrow *(Achillea millifolium)*	Seedling				
	Adult				
Flower border/field crop: Speedwells *(Veronica spp.)*	Seedling				
	Adult				
Established trees or shrubs: broad-leaved dock *(Rumex obtusifolius)*	Seedling				
	Adult				
Hard landscaped area: Groundsel *(Senecio vulgaris)*	Seedling				
	Adult				

Conclusions

1. List four types of herbicides available to control some common weeds.
2. Name one translocated and one contact herbicide.
3. State two ways in which selectivity of herbicide takes place.
4. State two conditions necessary for the successful use of each of the following herbicides:
    (a) contact
    (b) translocated
    (c) residual.
5. What legislation governs the use of herbicides in the UK?

Answers

Exercise 4.1. Weed collecting and identification

Conclusions

1. Any plant growing in the wrong place.

2. Produce a mass of seed. Rapid life cycle. Germinate early ahead of other plants. Dormant seeds can survive for many years in the soil.

3. Annual = sowthistle, scentless mayweed, sun spurge and red dead nettle. However, most ephemerals are also considered with annuals when studying weeds.

Perennial = creeping thistle, couch, yarrow and broad-leaved dock.

4. Competition for: light, water, nutrients, etc. Harbour pest and disease. Exude chemicals that inhibit other plants growing.

5. Clear associations exist between habitat and weed species found, e.g. yarrow grows in lawns, groundsel in borders, bindweed with shrubs, lichen on hard landscapes and *Oxalis* in glasshouses.

Exercise 4.2. Pressing and mounting weed specimens

Conclusions

1. To display as much as possible of the weed's main characteristics.

2. Incorrect pressing resulting in water still being present on the specimen which enables fungal growth to occur.

3. Risk of fungal growth due to incorrect pressing. It is important to change the tissue paper regularly to ensure all moisture is removed.

Exercise 4.3. Root structure (refer to Figure 4.3)

Conclusions

1. Fibrous because of the shallow rooting depth.

2. Tap root system because the deeper rooting depth enables increased soil water reserves to be accessed.

3. They need to be potted on into larger pots before they become pot-bound, otherwise their life span may be greatly reduced.

4. Suitability to the environment is reflected in the root system, e.g. mangrove roots provide support

holding the plant out of the water. The adventitious roots of ivy enable it to climb etc.

5. The most lignified would be stiff and harder to wriggle about.

6. Monocotledons mostly have fibrous roots; dicotyledons tap roots.

7. They are able to colonize larger areas, e.g. willow herb; often they propagate by both seed and vegetative means; cut short lengths are often able to regrow, e.g. couch grass; strong stem structure e.g. thistles.

Exercise 4.4. Cultural weed control

Conclusions

1. Because docks can regenerate even from small pieces of chopped tissue.

2. Results will vary between experiments. Generally the smaller the better.

3. Extremely successful.

4. Successful, however regrowth from underground tissue will be likely and regular recultivation is needed to keep the weed in check.

5. The grower might use a stale seed bed technique to induce weeds to grow then spray them off, plant out crops and regularly cultivate between rows to keep new weed growth under control.

6. Any four techniques listed in Background to Exercise 4.4.

Exercise 4.5. Chemical weed contol

Check current edition of *The UK Pesticide Guide* for details of chemical controls.

Conclusions

1.   Contact, selective, non-selective, systemic, residual, non-residual, soil acting or foliage acting.

2.   Translocated, e.g. glyphosate; contact, e.g. paraquat.

3.   (a) differential toxicity; (b) hormone toxicity.

4.   (a) Contact: clearly identified target weed; protection barrier for non target plants.
      (b) Translocated: dry conditions enabling absorption of chemical into the leaves; patience to wait several weeks for chemical to travel down to the roots before rotavating.
      (c) Residual: ensure chemical will not affect desired plants; wait the stipulated time interval until replanting.

5.   The Food and Environmental Protection Act 1985.

Chapter 5 The leaf and photosynthesis

Key facts

1. The rate of photosynthesis is influenced by environmental factors, many of which can be managed by the horticulturalist.
2. The law of the limiting factor dictates how to increase the rate of photosynthesis.
3. C3 and C4 plants have an important role to play in climate change and as sinks for carbon.
4. The starch test is used to test for photosynthesis.
5. The leaf structure and leaf area index are the primary influences on the rate of photosynthesis.
6. Chlorophyll is the green colour in leaves and stems but also comes in yellow, blue, orange, red and purple.
7. Photoperiodism regulates flowering period.
8. Plants may be short-day, long-day or day-neutral.
9. Many plants have adapted leaves to reduce transpiration.

Background

Photosynthesis is the process enabling plants to make food, in the form of carbohydrates (sugars and starch), using light energy.

Sugars are manufactured in the leaf from carbon dioxide (CO_2) and water (H_2O), using the energy of sunlight trapped by chlorophyll and other pigments. Oxygen (O_2), is the waste product given off by the plant. The sugars are transported around the plant (via the phloem) and deposited as starch. For most plants photosynthesis can occur over the temperature range 5–30°C. The optimum is considered to be 20–25°C, depending on species.

The rate of photosynthesis is influenced by the environment and as horticulturalists we can regulate the rate of photosynthesis by altering the supply of these factors. These include:

- Leaf area and age – managed by de-leafing, stimulating new growth
- Chlorophyll distribution – managed by de-leafing, stimulating new growth
- Light – managed by supplementary lighting, blackouts and shading
- Carbon dioxide levels – managed by CO_2 enrichment
- Temperature – managed by heating, ventilation
- Water supply – managed by irrigation, drainage
- Nutrients – managed by fertilizers, liquid feeding.

The **Law of the Limiting Factor** states that the factor in the least supply will limit the rate of process. We cannot increase photosynthesis by increasing a factor already in adequate supply. We must first identify which is the lowest factor and increase this. For example, we cannot increase carbon dioxide levels continually and expect to see an increase in growth without increasing other photosynthesis factors, such as temperature, as well.

The larger the leaf surface area, the faster will be the rate of photosynthesis and the subsequent growth and development of the plant. The leaf is one of the four primary organs of the plant. It may be defined as a lateral organ to the stem or axis of a plant below its growing point.

The leaf has four main functions. Most of these relate to the growth and development of plants through the processes of photosynthesis:

- manufacture food by photosynthesis
- enabling the diffusion of gases from leaf to atmosphere
- transpiration and cooling
- with modifications, as organs of food storage and vegetative propagation.

If we can therefore understand the structure and workings of the leaf we can increase the growth rate and ease of establishment of plants.

Different plant species have become adapted and specialized to their natural habitats around the world. Plant organs have adapted to most

The leaf is the main site for photosynthesis.

efficiently survive that environment and, even in a garden situation, the correct positioning of plants to take advantage of these features is important.

> Some leaves are adapted to aid **support**, e.g. sweet pea tendril, which is sensitive to touch (haptotropism) and twists around objects to support climbing.

Figure 5.1 Sweet pea tendrils

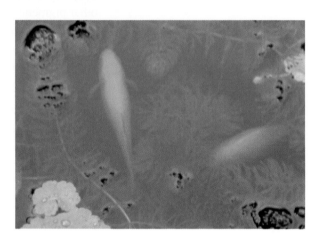

Figure 5.2 Pond weed are hydrophytes with extra air spaces in their leaves

Plants that live in water are called **hydrophytes** and have extra air spaces in their cells to trap oxygen (e.g. water lily, pondweed, duck weed), and subsequently flop when removed from their environment. By comparison **succulent** leaves are fleshy with their own water supply (e.g. echeveria, kalanchoe). **Carnivorous** plants are suitable for sites with poor nutrition and contain leaves modified to trap and digest insects (e.g. *Dionaea* (venus fly trap), sundew, butterwort and pitcher plants).

Starch test procedure

If photosynthesis is taking place the leaves should be producing sugars. Sugars are soluble materials that can be transported through the plant and converted into insoluble materials called starch. However, in many leaves, as fast as sugar is produced it is turned into starch. Since it is easier to test for the presence of starch than sugar, we can regard the production of starch as evidence that photosynthesis has taken place. The word and chemical formulas for the photosynthetic reaction are:

$$\text{Carbon dioxide} + \text{Water} \xrightarrow[\text{Chlorophyll}]{\text{Light energy}} \text{Sugar} + \text{Oxygen}$$

or

$$6CO_2 + 6H_2O \xrightarrow[\text{Chlorophyll}]{\text{Light energy}} C_6H_{12}O_6 + 6O_2$$

Thus, carbon dioxide and water are changed by chlorophyll using light energy into sugar and oxygen. The following exercises are designed to test this theory.

Aim

To test for the activity of photosynthesis (follow this procedure in Exercises 5.2–5.5).

Apparatus

Starch testing apparatus tray containing:

Test tube rack	Boiling tube (100 ml)	Pyrex beaker
Gauze mat	Forceps	Pencil
2 stirring rods	Iodine	Matches
Bunsen burner	Tripod	Dropping pipette
Industrial methylated spirit (IMS)	Heat tile	

Useful websites

www.footprints-science.co.uk/Starch.htm

www.chemsoc.org/networks/learnnet/cfb/photosynthesis.htm

Please note the following hazards

CAUTION

During these exercises you will be working with industrial methylated spirit (IMS). IMS is **highly flammable**. Do not use near an open flame. IMS has an **irritating vapour**. Do not inhale. If so, move into the fresh air. IMS is **harmful to the skin and eyes**. Wear eye goggles and take care not to spill any. If so, wash with plenty of water under a running tap.

Method

Set out the apparatus as in Figure 5.3. You may find it helpful to draw a labelled diagram of the apparatus.

1. Three quarters fill the pyrex beaker with water and place on the gauze mat, on a tripod over the Bunsen burner.

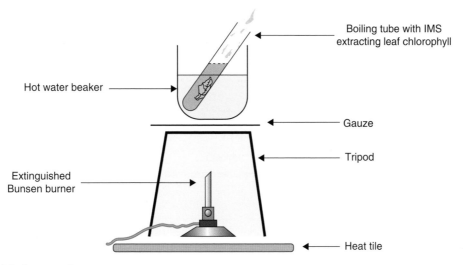

Figure 5.3 Starch testing apparatus

2. Light the Bunsen burner and heat the water until it boils.
3. Remove a leaf from the plant and boil in the water for 2 minutes. This ruptures the cell walls and prevents any further chemical reactions occurring.
4. Put on safety goggles and using the forceps remove the leaf and place into the test tube of IMS.
5. Turn off the gas to the Bunsen burner.
6. Place the test tube into the water bath and allow to warm until the leaf turns completely white (usually 5–10 minutes). This indicates that all the chlorophyll has been removed and makes the reaction of iodine with starch more clearly visible. The leaf will now be brittle and hard. IMS is dangerous and not to be heated directly. Always ensure that the tube is pointing away from you.
7. To return the leaf softness dip it again into the warm water bath.
8. Place the leaf onto a white tile and drench with iodine.
9. Observe the reaction over the next 2 minutes and record your results in the tables provided.

Results

Positive:	The reaction of starch (white), in the presence of iodine (yellow), is to change the leaf colour to deep blue or purple/black.
Negative:	If no starch is present then the iodine will stain the leaf yellow or brown.

Exercise 5.1

Dicotyledonous leaf structure and photosynthesis

Background

What is it about leaves that make them so important for photosynthesis and growth and development?

In typical leaf structure diagrams, adaptations to specific environments are described through leaf studies (e.g. margins, shapes and apexes). The amazon palm, for example, has a leaf blade 20 m long! The British broom, by comparison has tiny leaves. They may be either broad leaves (typical of dicotyledons) or narrow leaves (typical of monocotyledons). They often have distinctive textures such as small hairs which trap air to keep the plant cool (e.g. *Stachys lanata*). Similarly, *Erica carnea*, *Calluna vulgaris* and rosemary are adapted to dry conditions and save water by curling their leaves, creating a humid atmosphere.

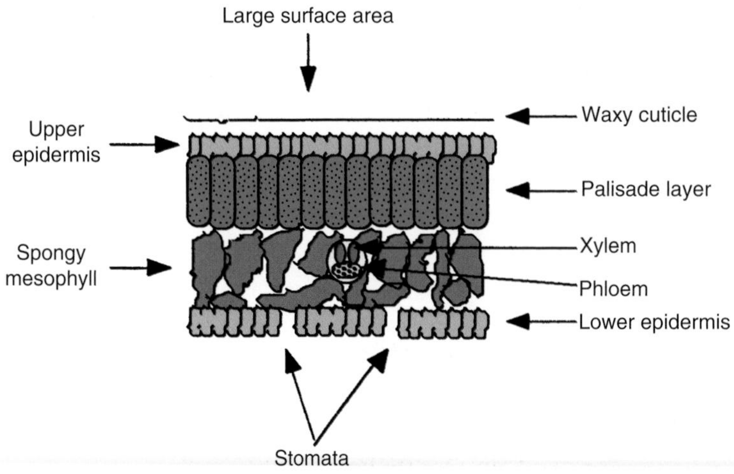

Figure 5.4 shows that there are several layers of different types of leaf tissue. These layers all have specialist jobs to perform to maximize the growth and development of the plant. Leaves have a **large surface**

Figure 5.4 Schematic representation of leaf structure

area onto which sunlight can fall. A waterproof waxy layer of **cuticle** covers the epidermis and is thicker on the upper surface of the leaf giving it a shiny appearance (e.g. camellia). Each time leaves are polished a layer of cuticle is removed and house plants are then more susceptible to wilting and disease.

The **epidermis** contains tightly fitting cells which help to keep the leaf air tight, protecting the leaf from water loss and bacteria and fungal penetration which might cause diseases. The epidermal layer is transparent, allowing light to pass through to the mesophyll layers below. This layer contains the chlorophyll and is arranged into two layers of tissue. First, the **palisade layer**. This is a layer of closely packed cells containing the green pigment called **chlorophyll**. Chlorophyll produces the green colour in plants and are the sites where photosynthesis (the manufacture of sugar) actually takes place. Second, the **spongy mesophyll**. This layer of loosely packed irregular shaped cells has many air spaces, allowing gaseous exchange between the leaf and the atmosphere. This enables carbon dioxide to get to the palisade layer and waste gases and moisture to escape through the stomata.

The veins (called vascular bundles) of leaves contain xylem and phloem transport vessels. Water and minerals are supplied to the leaf through the xylem. The phloem vessel transports the manufactured sugars, formed during photosynthesis, away from the leaf.

The **lower epidermis** contains small pore-like openings called **stomata.** These are pores in the underside of leaves. They are the entry and exit points for the diffusion of gases and water. The opening and closing of the stomata is controlled by the water content of the surrounding guard cells.

Leaf area index

The leaf area index (LAI) is the ratio of the total leaf area of the plant to the ground area covered by the plant. If a plant had just one layer of leaves each joining each other, then the leaf area would match the ground area and then the LAI would be 1.0.

On average LAI ranges from 2.0 to 5.0. Grassland LAI ranges from 0.3 to 2.0. Beech trees LAI averages 1.1, whereas LAI for conifer trees averages 0.5. Growing plants closer together, or indeed thinning plants out, greatly influences the LAI.

LAI is an important influence on photosynthesis and growth. The LAI influences sunlight absorbed, rainfall interception, energy conversion and gaseous exchange between the plant and the environment. The LAI for deciduous trees varies greatly over the seasons but changes little in conifers. Old leaves do not influence photosynthesis but do still help to catch rainfall.

Aim

To relate the functions of leaves to their component parts.

Apparatus

    Binocular microscopes
    Transitional section dicotyledon leaf slide

Useful websites

www.hillstrath.on.ca/moffatt/bio3a/plantphys/dicotleaf.html

www.biotopics.co.uk/plants/leafst.html

Method

Observe the specimen slides under the microscope.

Results

Draw a labelled diagram, in the space below, including the following terms:

waxy cuticle	upper epidermis	lower epidermis
palisade layer	lateral veins	spongy mesophyll
Stomata	vascular bundle	air spaces
main vein		

Conclusions

1. State the function of plant leaves.
2. Explain how you would select a plant suitable for a shallow, dry soil.
3. What leaf features would make a plant suitable for a sunny, exposed location?
4. Where is the leaf tissue found that is responsible for photosynthesis?
5. Explain what is meant by the term chlorophyll.
6. How do leaves reduce the risk of fungi and bacterial infection?
7. Name two of the tissues in the leaf which play a role in photosynthesis and state their particular function.
8. State four factors which effect the rate of photosynthesis.
9. Describe, with the aid of a diagram, how the structure of a dicotyledon leaf is designed to facilitate photosynthesis.
10. Explain the 'Law of the Limiting Factor'.

Exercise 5.2

Chlorophyll and photosynthesis

Background

Chloroplasts contain chlorophyll. Chloropyll is the green colour in plant leaves. This is the main plant pigment able to capture light energy and use it to turn carbon dioxide and water into sugar and oxygen. It is a mixture of pigments including chlorophyll-a (blue/green leaf colours), chlorophyll-b (yellow/green colours), carotene (orange leaf colours), xanthophyll (yellow autumn colours) and authocyanne (red/purple leaf colours). Chlorophyll is mainly found in the palisade layer of leaves. Many plants withdraw chloropyll from their leaves prior to leaf fall; as they do so the leaves show off autumn colours which are the various colours of chloropyll each being withdrawn in turn.

Figure 5.5 In cacti the leaves are reduced to spines and the stems turn green and photosynthesize

Variegated plants contain chlorophyll only in their green parts; only these areas should contain starch. However, in deserts the leaves of cacti are reduced to spines. The stem contains chlorophyll and is able to photosynthesize.

Chlorophyll production is partly dependent on nutrient supply and can be greatly influenced by our management of soil pH and fertilizer use. For example, a high soil pH (through over-liming) causes iron and manganese deficiency. This creates yellowing on the youngest leaves due to an absence of chlorophyll formation. In addition, over-fertilizing with potassium causes magnesium deficiency, resulting in yellowing of the middle and older leaves and an absence of chlorophyll production.

Aims

To test whether or not chlorophyll is necessary for photosynthesis by conducting a starch test on a variegated plant leaf and observing the locations of positive reactions.

Apparatus

    A variegated plant (e.g. geranium) labelled A
    Starch-testing apparatus tray

Useful website

www.footprints-science.co.uk/Starch.htm

Method

1. De-starch plants. At the beginning of each exercise, place the plants in darkness for a few days. In darkness the starch is changed into sugar and transported out of the leaf.
2. Remove Plant A from darkness and place in sunlight for several hours before beginning.
3. Remove a leaf from Plant A and draw a diagram of its shape and colours (green and white).
4. Conduct a starch test on the leaf as separately detailed and record your results in the table provided.
5. Draw a second diagram to show the subsequent colour change on reaction with iodine.

Results

The reaction of starch (white), in the presence of iodine (yellow), is to change the colour to deep blue or black. If no starch is present then the iodine will stain the leaf yellow or brown.

Variegated leaf colour	Colour reaction with iodine	Interpretation – starch present or absent
Green		
White		

Conclusions

1. What is chlorophyll?
2. What conclusions can be made about chlorophyll and photosynthesis?
3. Do variegated plants have more or less starch reserves?
4. Which vessel is the sugar solution transported away in?
5. Name the main pigment responsible for light absorption in photosynthesis.
6. Write out a word equation which summarizes the reactions in photosynthesis.

Exercise 5.3

Light and photosynthesis

Background

How light effects plant growth

Light has a variety of effects. The amount of light reaching the plant surface effects the rate of photosynthesis. In the shade, plants grow etiolated (tall, thin and yellow). This can effect the quality of the plant, e.g. *Aglaonaema* (foilage plant) which forms a better shape in moderate shade, rather than full light because the leaves grow flatter in shade. Also variegated plants such as *Dracaena* may form more variegation at 40,000 lux than at 20,000 lux and should therefore be kept in brighter light.

In horticulture light is measured in either lux or joules per square metre (J/m^2). A minimum of 500 lux is necessary for photosynthesis to take place. Optimum growth is achieved at 10,000–15,000 lux. The maximum amount of light that plants can usefully use is 30,000 lux. Interior landscape plants that prefer low light (400–600 lux) include *Aspidistra*, *Philodendron* and *Sansevieria*. Medium light (600–1000 lux) is preferred by *Anthriums*, *Cordyline* and *Scindapsus*. High light house plants (1000–2000 lux) include *Ficus benjamina*, *Maranta* and *Yucca*.

Photoperiodism

Different plant pigments absorb different wavelengths of light but mostly within the visible range of the electromagnetic spectrum. The colours most easily absorbed by plants are blue (400 nm), green (500 nm), yellow (600 nm), orange (650 nm) and red (700 nm). Often day length will act as a trigger to flowering. This is called photoperiodism.

Short-day plants which flower as the nights get longer include *Chrysanthemum*, *Poinsettia*, *Kalanchoe*, *Gardenia* and strawberry. **Long-day** plants which flower as the nights get shorter include *Petunia*, carnation, snapdragon, cabbage and tuberous *Begonia*. **Day-neutral** plants flower regardless of day length; examples include *Impateins,* tomato, cotton and *Fuchsia*.

The formula for photosynthesis implies that light is necessary for photosynthesis to occur. We regard the production of starch as evidence that photosynthesis has taken place. If we increase either the amount of light or provide light of greater intensity (brightness), the more photosynthesis can take place. There are several techniques that professional horticulturalists may use to achieve this, including:

Figure 5.6 Chrysanthemum are short day plants

- artificial light
- use of reflective surfaces (white wash)
- shading
- blackouts
- orientation of glasshouse or garden to south facing
- increased plant spacing
- reduced plant spacing
- remove weeds/eliminate competition
- clean glasshouse panes.

Figure 5.7 Carnations are long-day plants

Figure 5.8 Fuchsia are day neutral plants

Aim

To assess if plants can photosynthesize in both light and dark conditions.

Useful website

www.footprints-science.co.uk/Starch.htm

Apparatus

Geranium plants labelled B and C
Starch testing apparatus

Method

1. De-starch plants B and C before the exercise (see Exercise 5.2 for procedure).
2. Place plant B in sunlight for several hours. Keep plant C in darkness.
3. Remove a leaf from both plants B and C.
4. Conduct a starch test as detailed in the chapter background text.

Results

The reaction of starch (white), in the presence of iodine (yellow), is to change the colour to deep blue. If no starch is present then the iodine will stain the leaf yellow or brown.

Light conditions	Colour reaction with iodine	Interpretation – starch present or absent
Plant B (light)		
Plant C (dark)		

Conclusions

1. Which plant leaf went blue in the presence of iodine?
2. What can you conclude about the leaf that was kept in darkness?
3. What can you conclude about photosynthesis and the requirement for light?
4. Explain how light influences plant growth and development.
5. Describe the effect of low or no light on plant growth.
6. Explain why plant growth rate is lower in cloudy weather.
7. Describe how light levels may be manipulated to increase the rate of photosynthesis.

Exercise 5.4

Carbon dioxide and photosynthesis

Background

Carbon dioxide (CO_2) is believed to be a raw material in the process of photosynthesis.

Figure 5.9 Tomato plants often benefit from CO$_2$ enrichment

Can plants photosynthesize in a carbon dioxide free atmosphere?

Carbon dioxide comes from the air and enters the plant leaves through the stomata. The natural concentration of carbon dioxide in the air is 0.035 per cent (350 ppm). In unventilated glasshouses, plants use up carbon dioxide and the subsequent reduction in supply causes the rate of photosynthesis to slow down. By carbon dioxide enrichment in glasshouses, this concentration can be raised three times to 0.1 per cent (1000 ppm), leading to an increase in the rate of photosynthesis and increased plant growth.

Enrichment often occurs during the winter months on sunny days when an enriched atmosphere can be maintained because the vents are closed. An injection rate of 15–55 kg/ha/hr (kilograms per hectare per hour) under moderate wind conditions is normal. In the summer months, when crops are growing vigorously, carbon dioxide levels in the crop may fall to 0.01 per cent (100 ppm) and even with fully open vents injection rates of 25–30 kg/ha/hr are required to maintain concentrations at normal atmosphere levels (0.035 per cent or 350 ppm). Economic returns vary with the size of glasshouse and other variables, but as a general rule for every 1p spent on carbon dioxide enrichment, a return of 4p is yielded through increased crop yield.

C3 and C4 plants

It is possible to divide most plants into two categories (C3 and C4), based upon how they assimilate carbon dioxide. This happens in the mesophyll cells of the leaf. About 95 per cent of plants are in the C3 group. C3 plants are so called because during CO$_2$ assimilation they initially form carbon molecules made from three carbon atoms. C3 plants are more efficient than C4 plants under cool, wet conditions and under normal light. C3 plants are mainly woody and round leaved, and include sugar, barley, beet, oats, wheat, rice, cotton, tomatoes, potato, peanuts, trees, sunflowers and most temperate plants.

C4 plants, by comparison, initially form carbon molecules made from four carbon atoms. They photosynthesize faster than C3 plants under high light levels and high temperatures. Thus photosynthesis is quicker and more efficient. They make better use of water and consequently lose less water to transpiration (about 25 per cent of C3), because they do not need to keep the stomata open as long for the same volume of CO$_2$ absorbed. For this reason they are found in drier, hotter climates. C4 plants come from at least 19 families and include thousands of species such as corn, sugar cane, maize, millet, grasses and sedges.

One important significance of this difference is that as CO$_2$ levels rise in the environment, C3 plants are able to increase their rate of photosynthesis without dumping extra CO$_2$ into the atmosphere. C4 plants cannot respond by increasing photosynthesis.

Aims

To sustain a plant in a carbon dioxide free environment and observe whether or not photosynthesis is possible by monitoring the presence of starch in the plant leaves.

Apparatus

Set out the apparatus as in Figure 5.10.

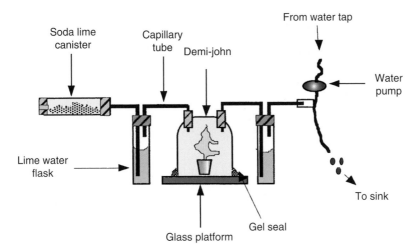

Figure 5.10 Carbon dioxide extracting apparatus

Geranium plant labelled D	Soda lime canister	Lime water flasks × 2
Capillary tubes × 5	Demi john chamber	Water pump
Glass platform	High vacuum gel	Starch testing apparatus

Method

You may find it helpful to draw a labelled diagram of the apparatus.

1. De-starch plant D, and seal into the carbon dioxide free chamber for several hours (see Exercise 5.2 for procedure).
2. Remove a leaf from plant D.
3. Conduct a starch test as separately detailed and record your results below.

Results

The reaction of starch (white), in the presence of iodine (yellow), is to change the colour to deep blue. If no starch is present then the iodine will stain the leaf yellow or brown.

CO_2 free environment	Colour reaction with iodine	Interpretation – starch present or absent
Plant D		

Conclusions

1. Why is soda lime used?
2. Explain why lime water flasks are used.
3. How does carbon dioxide enter the plant?
4. What can you conclude about carbon dioxide and photosynthesis?
5. Explain why a lack of carbon dioxide reduces photosynthesis.
6. Describe how carbon dioxide levels may be manipulated to increase the rate of photosynthesis.

Exercise 5.5

Water and photosynthesis

Background

> Water is important to plants in many ways including maintaining leaf structure for photosynthesis.

Water is supplied to the leaf through the xylem vessel. Although only 1 per cent of water absorbed by plants is used for photosynthesis, any lack of water in the leaves causes water stress. Water is essential to maintain leaf turgidity (cell strength), without which it would not be possible to open the stomata and allow carbon dioxide into the leaf. A plant that is wilting cannot therefore photosynthesize, and if allowed to continue to wilt would result in plant death.

Some plants are adapted to prevent water loss. For example in gorse, broom and horsetails the leaves are reduced to scales, lowering the surface area and reducing water loss. The stems contain chlorophyll and photosynthesize. There are several techniques that professional horticulturalists may use to prevent water loss. Here are some examples of both plant and horticulturalist strategies.

Plant strategies
- thick cuticles
- close stomata
- reduce surface area
- create a humid atmosphere e.g. curl leaves.

Horticulturalist's strategies
- shading
- increase humidity e.g. fogging, mist, polythene
- lower the air temperature
- reduce leaf area
- reduce air movements e.g. wind breaks
- spray with anti-transpirants.

Figure 5.11 Overhead irrigation in nursery stock

Aim

To observe if plants can photosynthesize in the absence of water.

Apparatus

Geranium plant labelled E
Starch test apparatus

Useful website

www.footprints-science.co.uk/Starch.htm

Method

1. De-starch plant E before the exercise.
2. Do not water plant E for several days before the exercise.
3. Remove a leaf from plant E.
4. Conduct a starch test as separately detailed and record your results in the table provided.

Results

The reaction of starch (white), in the presence of iodine (yellow), is to change the colour to deep blue. If no starch is present then the iodine will stain the leaf yellow or brown.

Water-free environment	Colour reaction with iodine	Interpretation – starch present or absent
Plant E		

Conclusions

1. What can you conclude about water and photosynthesis?
2. How does the water enter the plant for photosynthesis?
3. How many units of water are required in photosynthesis?
4. List two ways in which plant leaves are adapted to reduce water loss.
5. Write out a chemical equation which summarizes the reactions in photosynthesis.

Exercise 5.6

Oxygen and photosynthesis

Background

Photosynthesis results in the production of sugar, which is stored as starch and used later for growth and development (respiration), and oxygen, which is a waste product.

Aims

To observe the production of oxygen by pond weed under light and dark conditions.

Apparatus

Plants labelled F and G (pond weed)
Beakers $\times 2$

Glass funnel

Test tube

Oxygen meter

Method

You may find it helpful to draw a labelled diagram of the apparatus.

1. Keep the plants in a beaker of water throughout the exercise covered by a glass funnel.
2. De-starch plants F and G before the exercise (see Exercise 5.2 for procedure).
3. Keep plant F in darkness. Place plant G in the light for several hours.
4. Observe the gas being produced under each condition.
5. Record the oxygen content of the water using the oxygen meter.

Results

Enter your results in the table provided.

Light/dark environment	Oxygen level recorded	Interpretation
Plant F (light)		
Plant G (dark)		

Conclusions

1. What can you conclude about oxygen and photosynthesis?
2. How does the waste oxygen escape from the plant?
3. Complete the gaps in the following text:

For a plant to produce carbohydrates, it absorbs from the surrounding air _____ and _____ in the form of gases. These enter the plant through the _____ on the leaves. The carbohydrates produced by the plant are used as _____ in the process of growth. Water is also supplied through the _____ vessels within the leaf veins (vascular bundles).

4. List the ingredients of photosynthesis under the following headings:
   Consumable inputs
   Resulting products
   Processes systems.
5. List five factors which effect the rate of photosynthesis.
6. List the methods which a horticulturalist can use to maximize the rate of photosynthesis during the winter months.

Exercise 5.7

Cells and photosynthesis

Background

The basic unit of plant structure is the cell. It is composed of a nucleus within a mass of protoplasm bounded by a membrane (see Figure 5.12). Energy for cell growth and specialization comes from sugar made during photosynthesis in the leaf.

These cell parts (organelles) are living and continually move. The simplest cells have the ability to divide to form two identical cells. All cells begin life

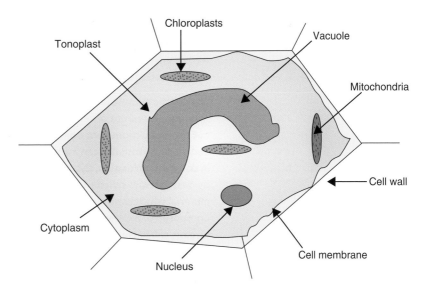

Figure 5.12 Cell structure

unspecialized and may then develop under the control of genetic information, into specialized cells.

Unspecialized cells are called **parenchyma.** These are present in most of the plant as packing cells, e.g. leaf mesophyll and the soft flesh of fruit. They are capable of division and forming new plant parts, e.g. new roots when taking cuttings. They are larger than meristematic cells, vacuolated and may contain chloroplasts, e.g. leaf pallisade cells. They are capable of cell division and of forming new cells if the plant is wounded. Ninety per cent of plant growth occurs by vacuolation, causing cells to enlarge.

Parenchyma are usually loosely packed with air spaces present between the cells useful for gaseous exchange. The cells absorb water by osmosis until they are turgid, enabling stems to remain erect. If they lose water and become flaccid, the plant wilts.

Tissues where cell division occurs and where cells are relatively unspecialized are called meristematic. They have thin walls, a large nucleus and are full of cytoplasm. There are no vacuoles. These cells cause growth by dividing by mitosis. This is vital for secondary thickening of woody plants. Stem elongation in woody plants can be very seasonal, perhaps confined to a few weeks in the spring. The rest of the growing season may be spent in forming next year's bud. Examples of meristematic cells include in root tips, shoot tips and the cambium.

There are several types of specialist cells such as epidermal cells which are tightly packed for protection. In later years these may be replaced by bark. The epidermis secretes a thin waxy layer called cuticle, which waterproofs the plant by over ninety per cent and also prevents fungal spore penetration. **Collenchyma** cells have cell walls thickened with cellulose to provide strength. Cells become long and thin, e.g. celery, leaf epidermis and young stems. Lamiaceae family plants have extra ribs of collenchyma.

Sclerenchyma cells form together into fibres. The cells are dead and thickened with lignin laid down on the outside of the cellulose cell wall. They are often long and thin, and joined together in a string up the stem e.g. flax in rope, leaf epidermis, roots, shells of nuts, the grittiness in pear flesh, xylem and phloem vessels.

Cells are composed of organelles (see Figure 5.12). The **cell wall** is made of cellulose fibres giving strength and rigidity. The wall is completely permeable to water and nutrients. Closely packed cells offer better protected from aphids, which need to get in between cell walls. In xylem and phloem tissue the cell walls are waterproofed and strengthened with a completely impermeable material called lignin.

The **cell membrane is** semi-permeable and controls the uptake of water and nutrients in and out of cell. The **cytoplasm** contains the machinery for carrying out the instructions sent from the nucleus. It is a jelly-like material of protein and water. Cytoplasm can form as little as ten per cent of cell volume and is in constant circulation. Cytoplasm from one cell is connected to another cell by passageways called **plasmodesmata**. The **nucleus** contains the genetic material which controls the cell, regulating growth and reproduction. Composed of a chemical called deoxyribonucleic acid (DNA) which forms long threads called chromosomes only visible in cell division, it carries the gene code for the inherited characteristics of the plant. **Chloroplasts** contain the green pigment chlorophyll which is essential to trap light for photosynthesis. Mitochondria are the power house of the cell. They are tiny bodies that contain the enzymes responsible for respiration. Mitochondria contain tiny chemical motors rotating over 1000 times a second to produce ATP synthase, which is the central enzyme in energy conversion in mitochondria. The **vacuole** is a large, water filled, reservoir and waste disposal system that can occupy up to 85 per cent of the volume of the cell, helping to regulate water uptake by the plant. The **tonoplast** is the vacuole's membrane.

Aim
To consolidate understanding of cell tissue and function.

Apparatus
List of cell organelle components

Chloroplasts	Vacuole
Mitochondria	Cell wall
Cell membrane	Nucleus
Cytoplasm	Tonoplast

Method
Match the correct organelle to the functions listed.

Results
1. Site of photosynthesis:
2. Vacuole's membrane:
3. Strength and rigidity, holding everything together:
4. A jelly-like material of protein and water:
5. Control centre, regulating growth and cell reproduction:
6. Site of respiration:
7. Helps regulates water uptake by the plant:
8. Regulates the movement of substances in and out of the cell:

Conclusions

1. Name the tight-fitting cells surrounding the plant for protection.
2. State the name of unspecialized packing cells.
3. Name the specialized cell, thickened with cellulose.
4. Name specialized cell surrounded by lignin.
5. Roots, nut shells and gritty peach flesh are formed from these cells strengthened with this.
6. Strands of celery are examples of the cells formed with this compound.
7. Leaf palisade and mesophyll cells are example of these cells.

Answers

Exercise 5.1

Diagram similar to Figure 5.4 required.

Conclusions

1. To provide energy through photosynthesis

2. The plant would be likely to suffer from drought and therefore one with modified leaves to reduce moisture loss should be selected, e.g. rosemary.

3. High transpiration would be encountered so modified leaves such as those with waxy cuticles or leaf curls/hairs should be used, e.g. heathers.

4. Chloroplasts are located within the palisade layer.

5. The green colour in leaves and stems that is the site of photosynthesis.

6. Tightly packed epidermal cells and waxy cuticles.

7. Any two tissues named and discussed as in Background section.

8. Any four from leaf area and age, chlorophyll distribution, light, temperature, water supply or nutrients.

9. Diagram as Figure 5.4. All tissues labelled and their function described.

10. The factor in least supply will limit the rate of the process. We cannot continue to increase a factor, e.g. light levels, without addressing all the other ingredients of photosynthesis.

Exercise 5.2

Only the green parts will contain starch, suggesting that chlorophyll is necessary for photosythesis to occur.

Conclusions

1. The green colour in leaves and stems that is the site of photosynthesis.

2. Chlorophyll is essential for photosynthesis.

3. Less.

4. Phloem tissue.

5. Chlorophyll.

6. Carbon dioxide + water in the presence of light and chlorophyll change into sugar and waste oxygen.

Exercise 5.3

Conclusions

1. Plant B. The plant kept in the light.

2. It was unable to photosynthesize and produce starch.

3. Light is essential for photosynthesis to occur.

4. Light affects both growth and development, e.g. etiolation and photoperiodism with short, long and day-neutral plants

5. Plants grow etiolated (long, thin and stretched).

6. Choice from artificial light, reflective surfaces, shading, blackouts, orientation, plant spacing, removing weeds and cleaning windows.

Exercise 5.4

Plant D remained yellow.

Conclusions

1. Soda lime removes carbon dioxide from the atmosphere.

2. Lime water reacts with carbon dioxide to turn the water milky. Thus the bottles indicate whether or not carbon dioxide has been successfully stripped from the air.

3. Through the leaf stomata by a process called diffusion.

4. Carbon dioxide is essential for photosythesis to occur.

5. The Law of the Limiting Factor applies and photosythesis slows.

6. By using enrichment in glasshouses to triple the normal concentration.

Exercise 5.5

Plant E remained yellow.

Conclusions

1. Water is essential for photosynthesis to occur.

2. Through the roots by a process called osmosis.

Exercise 5.6

Pond weed kept in the light produced substantially more oxygenated water.

Conclusions

1. Oxygen is a waste product from photosynthesis.

2. Through the leaf stomata via a processes called diffusion.

3. Carbon dioxide, stomata, energy, xylem.

Exercise 5.7

1. Chloroplasts

2. Tonoplast

3. Cell wall

4. Cytoplasm

5. Nucleus

6. Mitochondria

7. Vacuole

8. Cell membrane

3. Six.

4. Leaf hairs, curls, waxy cuticles or leaves reduced to spines

5. $6\,CO_2 + 6\,H_2O \rightarrow$ light/chlorophyll $\rightarrow C_6H_{12}O_6 + 6O_2$

4. Inputs: carbon dioxide and water
   outputs: sugar and oxygen
   systems: light and chlorophyll

5. Choice from: leaf area and age, chlorophyll distribution, light, temperature, water supply or nutrients.

6. Choice from: artificial light, reflective surfaces, orientation, plant spacing, removing weeds, cleaning windows or carbon dioxide enrichment.

Conclusions

1. Epidermis

2. Parenchyma

3. Collenchyma

4. Sclerenchyma

5. Lignin

6. Cellulose

7. Parenchyma

Chapter 6 Respiration and storage

Key facts

1. Respiration is the process through which plants burn energy for life processes.
2. Aerobic respiration occurs in the presence of air and produces far more energy than anaerobic respiration (without air).
3. During respiration plants release carbon dioxide and help with the carbon cycle.
4. The carbon cycle helps us to understand the role plants have in sustainable development and global warming.
5. Horticulturalists can manage respiration rates so as to encourage cuttings to root, or slow it down to extend the shelf-life of flowers, fruit and vegetables.
6. Respiration rates in plants rises after attack by animals or insects as the plant repairs itself.

Background

Respiration is the destruction (use) of sugars and starch, made during photosynthesis, to release energy for the plant's life sustaining processes.

Figure 6.1 Chrysanthemum cut flowers

All organisms need to respire

When plants are harvested for selling, respiration continues. Plant sugars are being used up without being replaced (as they are not photosynthesized) and the produce, be they flowers, vegetables or fruit, will eventually break down and decompose. Cut flowers in particular are fragile and perishable materials. Prosperity in the expanding trade in cut flowers is dependent on a regular supply of high-quality, fresh flowers with a long vase life. They will eventually die, but with proper care and attention, this can be delayed. Flowers and plants that do not last as long as expected create dissatisfied customers.

Respiration is essential for plant growth and development. The liberated energy is used for cell division and making useful plant substances, including cellulose in cell walls, proteins, enzymes asnd general growth and repair.

Managing plant respiration enables us to hold larger stocks that will not perish so easily and the customer obtains greater value for money.

Figure 6.2 Shelf life trial in process

Figure 6.2 shows a typical shelf life trial in progress. Similarly, pot plants should not be stood in draughts, direct sunlight or where it is too hot. Water when required. Remove spent blooms and dead leaves. Use leaf shiners and protective wrapping to reduce respiration rates.

The process of respiration takes place in all living cells in the plant. It occurs most rapidly at the meristematic points (growth points) where cells are actively dividing (e.g. roots and shoots). It is a continual process occurring during day and night.

Because energy is used for plant growth and development, respiration results in a decrease in plant sugar reserves. This is acceptable, if the sugars are being replaced through photosynthesis, but in certain circumstances this is not possible and plants may subsequently shrivel and die; for example wilting plants, cut flowers, and stored fruit and vegetables.

In these conditions attempts must be made to slow down respiration and extend the life of the plant. Respiration rates are influenced by a variety of environmental factors including:

- temperature – respiration rate of many flowers at 2°C is only 10 per cent of the rate at 20°C. In roses, the respiration rate at 25°C is six times faster than at 6°C. This suggest that one day at 25°C is equivalent to six days at 6°C.
- water
- oxygen level – reduced oxygen levels helps preserve fruit in colds stores
- plant injury (wounding or infection) – injury is often caused by rough poor handling and plants become susceptible to pest and disease attack.

Respiration may be either aerobic (with air) or anaerobic (without air).

Aerobic respiration

$$\text{Sugar} + \text{Oxygen} \rightarrow \text{Carbon Dioxide} + \text{Water} + \text{Energy}$$
$$C_6H_{12}O_6 + 6O_2 \rightarrow 6CO_2 + 6H_2O + 2830 \text{ kJ}$$

Anaerobic respiration

$$\text{Sugar} \rightarrow \text{Carbon Dioxide} + \text{Alcohol} + \text{Energy}$$
$$C_6H_{12}O_6 + 2CO_2 + 2C_2H_5OH + 118 \text{ kJ}$$

The exercises in this chapter are designed to investigate these ingredients and to ascertain the best storage method to reduce respiration, and therefore extend product shelf life.

The carbon cycle

Plants use sunlight to convert carbon dioxide to sugar during photosynthesis. Carbon in the atmosphere forms carbon dioxide (CO_2), which acts as a greenhouse gas. Carbon also exists dissolved in rain water, 'locked up' in rocks and in dead organic matter. For life to continue carbon must be recycled. During plant respiration CO_2 is a waste product, recycled back into the atmosphere, but the majority of CO_2 in the air is released from animals. It is also 'locked up' in living plant matter but released when it decays or is burnt. In addition, carbon is cycled through the soil, soil organisms, burning of fossil fuels and volcanic eruptions. By planting more trees and shrubs, horticulturalists can help to 'lock up' carbon that might otherwise contribute to global warming, within the plant structure.

Figure 6.3 Sustainable planting using trees and shrubs to lock in carbon

Exercise 6.1

Storage of plant materials

Background

While the plant is growing, respiration rates are needed that are high enough to provide the energy for the production of new structures, but not so high that excessive amounts of energy are lost in unnecessary vegetative growth. The grower must balance environmental factors such as temperature and light, with respiration, so that the sugars broken down by respiration do not exceed those produced by photosynthesis, otherwise there will be negative growth.

However, when plants are harvested for selling or display, respiration continues. Plant sugars are being used up without being replaced, as they are not photosynthesized, and the produce, be they flowers, vegetables or fruit, will eventually breakdown and decompose. There are several circumstances were it is desirable to slow down the rate of respiration to prevent decay, such as:

Figure 6.4 Keep cut flowers cool to extend shelf life

Night time	Respiration occurs during both day and night, but photosynthesis occurs only during the day. The plant must therefore store enough sugar during the day to fuel night-time respiration. Respiration increases with temperature. If we lower the temperature at night (by ventilation) then we can slow down the rate of respiration.
Storage	Respiration continues when plants are harvested. Sugars are being used up without being replaced, and the plant will eventually decay. We therefore seek to reduce respiration by putting the produce in cold stores (e.g. 0–10°C storage for cut flowers, apples, vegetables onions, cuttings, and chrysanthemums).

Figure 6.5 Apples stored in low oxygen stores can taste of alcohol

Storage

Similarly, respiration increases with oxygen levels. Therefore if we lower the oxygen supply in controlled atmosphere stores, decay will be reduced (e.g. cut flowers stored in a florist's shop). However, if oxygen levels are very low, aerobic respiration stops and anaerobic respiration starts. Alcohol is a waste produced and is toxic to plants. In apple storage, low oxygen levels, caused by carbon dioxide build-up, result in an alcohol flavour. In comparison, a high carbon dioxide level leads to brown heart in pears and core flush in apples. Table 6.1 shows some example storage periods for a range of plants.

Aim

To assess the effectiveness of different storage systems on reducing respiration rates in carrots and therefore extending the shelf life of our produce (cut flowers are also suitable for investigation).

Apparatus

6 carrots
Top-pan analytical balance
Incubating ovens
Cold storage fridge

Table 6.1 **Storage periods**

Plant	Cold storage period
Vegetables	
Spinach	1 week
Green beans	10–14 days
Cauliflower	3–5 weeks
Carrots	12 months
Potatoes	12 months
Fruits	
Strawberries	6–9 days
Raspberries	6–9 days
Blackberries	8–10 days
Damson plums	12–14 days
Cut flowers	
Tulip, lily, gerbera	2–4 weeks
Rose	1–2 weeks
Freesia, hyacynth	2–4 weeks
Chrysanthemum	3 weeks
Carnation	4–6 weeks

Plastic bags
Sample labels
Indelible pens

Method

1. Arrange for one carrot to be stored under each of the following environmental conditions:
   A high temperature incubator (30°C), sealed in a plastic bag
   B high temperature incubator (30°C), without a plastic bag
   C cold storage fridge (4°C), sealed in a plastic bag
   D cold storage fridge (4°C), without a plastic bag
   E ambient (room) temperature without a plastic bag
   F ambient (room) temperature in a plastic bag

2. For those carrots that are placed in a plastic bag, place a label on the carrot and weigh the carrot before placing in the bag.
3. After labelling, weigh your carrots before storing them, using the analytical balances and the following technique:
   (a) 'Tare' (zero) the balance. This sets the microprocessor memory to read zero.
   (b) While the microprocessor is doing this, the balance readout display will be blank. Do not put anything on the balance while this is happening.
   (c) When the display returns to '0.00 g' commence weighing your sample.
4. Record the masses in the table provided.
5. Label your samples, using the indelible pens, providing the following information:
   (a) name
   (b) date sample prepared
   (c) weight of carrots recorded
   (d) storage method A, B, C, D, E or F.

6. Leave your samples in storage for two weeks.
7. After two weeks:
   (a) remove your samples from storage and reweigh them
   (b) record this as the mass after storage, in the table provided.

Results

1. Calculate the difference in weight between before and after storage.
2. Calculate the percentage change in weight using the following formulae:

$$\text{Percentage change in mass} = \frac{\text{mass after storage} - \text{mass before storage}}{\text{mass before storage}} \times 100$$

Storage treatment method	Mass before storage (g)	Mass after storage (g)	Change in mass (g)	% change in mass
A				
B				
C				
D				
E				
F				

Conclusions

1. Explain your results in terms of respiration rates and storage treatments.
2. Rank your storage treatments in order of their efficiency at reducing respiration.
3. Explain the significance of respiration rate in relation to storage of harvested produce and seeds.
4. State two instances where horticulturists might want to slow down the rate of respiration.
5. Explain two ways in which a reduction in the rate of respiration might be achieved.

Exercise 6.2

Aerobic germination of pea seeds

Background

Aerobic respiration is the process whereby carbohydrates (sugars and starch) are broken down using oxygen to liberate energy, with the production of waste carbon dioxide and water.

Carbon dioxide, when mixed with lime water, will turn the water milky and is therefore an indicator that respiration has taken place. Without water respiration cannot take place (e.g. seeds will not germinate without water).

Aim

To demonstrate that germinating seeds respire and to identify the gas given off.

Apparatus

Water-soaked pea seed
Water-soaked boiled peas
Cotton wool
Corked conical flasks A and B
Test tube of lime water

Figure 6.6 shows the apparatus set up.

Method

You may find it helpful to draw a labelled diagram of the apparatus.

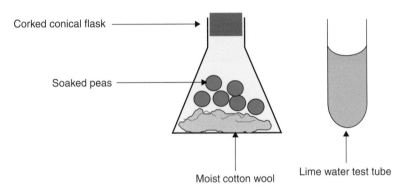

Corked conical flask

Soaked peas

Moist cotton wool

Lime water test tube

1. Place the living, water-soaked seeds in conical flask A.
2. Place the boiled peas in flask B.
3. Leave both flasks several days to germinate.
4. Uncork both flasks and tip the gas into the lime water. Carbon dioxide is heavier than air, and although colourless, will flow if the flasks are poured gently into the lime water.

Figure 6.6 Germinating peas and carbon dioxide testing apparatus

Results

1. Record what happens to the lime water in flasks A and B.
2. What is the gas being produced?
3. What process results in the production of this gas?
4. Why were some of the peas boiled?
5. What effect does boiling the seed have on germination ability?

Conclusions

1. What are your conclusions from this experiment?
2. Explain the meaning of the term 'aerobic respiration'?
3. State two factors which increase the rate of respiration.

Exercise 6.3

Anaerobic respiration of yeast

Background

Anaerobic respiration is respiration without oxygen. Alcohol is produced as a waste product and is toxic to plants. Anaerobic respiration produces less energy for the same amount of sugar than aerobic respiration (4 per cent) (from $118/2830 \times 100$).

$$\text{Glucose} \rightarrow \text{Carbon dioxide} + \text{Alcohol} + \text{Energy}$$
$$C_6H_{12}O_6 \rightarrow 2CO_2 + 2C_2H_5OH + 118 \text{ kJ}$$

There are several circumstances in horticulture were this occurs. Management action should be taken to prevent anaerobic conditions developing. Examples of where such environments occur include:

- poorly aerated soils – results in poor growth/germination
- poor film in hydroponic production – results in low oxygen and low growth
- modified atmosphere packaging – results in mould growth and alcohol flavour
- compaction – results in poor growth, sulphur dioxide gas (rotten egg smell)
- overwatering – results in root death, stunted growth.

Figure 6.7 Compacted soils often give off a rotten egg smell, when forked due to anaerobic respiration of roots

Aim

To demonstrate that anaerobic respiration can take place. This exercise designed so that the yeast is living in an anaerobic environment.

Apparatus

Test tube
Water bath
Live yeast
Capillary tube
Liquid paraffin
Lime water filled test tube

Figure 6.8 shows how to set up the test tubes.

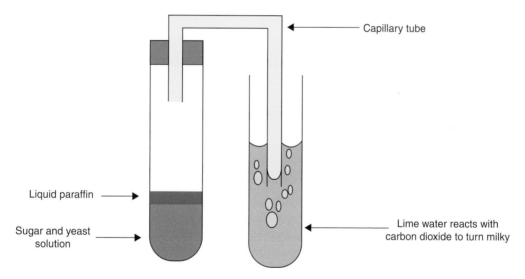

Liquid paraffin

Sugar and yeast solution

Capillary tube

Lime water reacts with carbon dioxide to turn milky

Figure 6.8 Anaerobic respiration apparatus

Useful websites

www.csun.edu/scied/2-longitudinal/schuster/index.html

Method

You may find it helpful to draw a labelled diagram of the apparatus.

1. Prepare de-oxygenated water by boiling distilled water.
2. To the test tube add 5 ml of glucose solution (5 per cent mass/volume or m/v), prepared using de-oxygenated water, and 1 ml of yeast suspension (10 per cent m/v), also prepared using de-oxygenated water.

3. Cover the mixture with a thin layer of liquid paraffin to exclude oxygen from the mixture.
4. Connect the test tube of lime water to the test tube of yeast, using a capillary tube.
5. Place the yeast mixture test tube in a warm water bath to stimulate growth.
6. Leave the mixture for 30 minutes and await a reaction.

Results

1. How long does it take before the lime water reacts?
2. What happens to the lime water?
3. What is the gas given off?
4. What process results in the production of this gas?
5. Why is distilled water used?
6. Why is deoxygenated water used?

Conclusions

1. Explain what is meant by 'anaerobic respiration'.
2. State two places where anaerobic respiration might occur in horticulture and state its importance.
3. Explain why anaerobic conditions are undesirable in horticulture.

Exercise 6.4

Energy release during respiration

Background

Heat being produced is a good indication that energy is being used up during respiration.

Respiration is a chemical reaction. Chemical reactions increase with temperature, therefore respiration increases with temperature. For every 10°C rise in temperature between 10°C and 30°C, the respiration rate will double. At temperatures above 30°C, excessive respiration occurs, which is harmful to the plant. At temperatures above 40°C, plant tissue collapses and breaks down.

Respiration rates may also be elevated when the plant has been injured in some way. Under these conditions the plant starts to heal itself through producing new cells to repair the damage. This new growth is fuelled by respiration. Any plant injury to tissue, either by mechanical damage or by infection, increases the rate of respiration. For example, damaged stems caused by overcrowding in a bucket of cut flowers results in sugars being used up as the plant repairs the damage.

However, there are times when we want to increase plant respiration rates to increase growth and development. The respiration rate at the base of the cuttings is increased after the cutting is taken by applying heat to the root zone (basal root zone warming). Similarly, seed respiration rates increase as they germinate. They need energy to break down cotyledon food store and produce new cell growth.

Aims

To investigate heat as a product of respiration during the germination of wheat seeds.

Apparatus

Vacuum flasks A and B
Cotton wool
1 litre of wheat seeds
Thermometers A and B
1 per cent formalin solution

Figure 6.9 shows how each flask is to be set out.

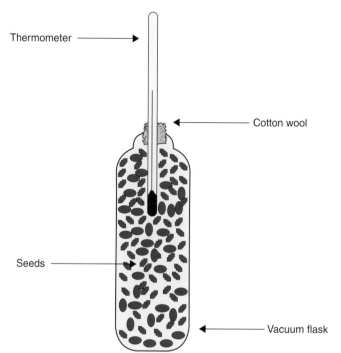

Figure 6.9 Energy release during respiration

Method

You may find it helpful to draw a labelled diagram of the apparatus.

1. Rinse the wheat seeds in 1 per cent formalin solution, to kill all bacteria and fungi on the grains.
2. Place half of the wheat seeds in vacuum flask A.
3. Boil the other half of the wheat seeds and place the cooled seed in vacuum flask B.
4. Place a laboratory thermometer in both flasks A and B.
5. Seal each flask with cotton wool and leave to germinate for a few days.
6. After a few days record the temperature of flasks A and B in the table below.

Results

Enter your results in the table provided.

Seed treatment	Recorded temperature (°C)
Flask A	
Flask B	
Temperature difference	

Conclusions

1. Which flask has the highest temperature?
2. Why are the flasks sealed with cotton wool?
3. What can be concluded about respiration from this experiment?
4. Describe how the natural healing ability of plants is exploited in plant propagation.
5. Explain the significance to the grower of the process of plant respiration in obtaining optimum temperatures for growth of a crop.
6. Compare the process of respiration with photosynthesis.

Answers

Exercise 6.1. Storage of plant material

Results will vary between experiments and operators.

Conclusions

1. Respiration is clearly linked to temperature. The combination of reduced temperature and a sealed environment creating a modified atmosphere was the most successful at reducing respiration and extending shelf life.

2. A = 5, B = 4, C = 1, D = 2, E = 4, F = 3.

3. All plants will continue to respire after harvest since it occurs in every cell 24 hours a day. However, the energy is not being replaced by photosynthesis and the plant will slowly decay. Since this is undesirable, respiration must be slowed by decreasing temperature, careful handling to avoid bruising plants and wrapping appropriately to create a modified atmosphere.

4. (1) To extend the shelf life of cut flowers and plants. (2) When taking cuttings to reduce growth of the stem before the new roots have taken.

5. Choice from: reduced temperature, misting/spraying, careful wrapping, or low oxygen storage.

Exercise 6.2. Aerobic germination of pea seeds

Results

1. Flask A lime water turned milky. Flask B saw no reaction.

2. Carbon dioxide.

3. Aerobic respiration.

4. As a control to see if dead peas also produced carbon dioxide.

5. They are dead and will no longer germinate.

Conclusions

1. Carbon dioxide is produced as a waste product of aerobic respiration.

2. Respiration in the presence of air.

3. Choice from: temperature, oxygen, wounding, bruising, pest and disease, draughts.

Exercise 6.3. Anaerobic respiration of yeast

Results

1. 15 minutes.

2. It turns milky.

3. Carbon dioxide.

4. Respiration.

5. It is the purest form of water available.

6. To ensure anaerobic conditions.

Conclusions

1. Respiration without air.

2. It is very important since alcohol is produced as a waste product and is toxic to the plant. Choice from: poorly aerated soils, compacted soils, overwatering, poor film in hydroponics and modified atmosphere stores/packaging.

Exercise 6.4. Energy release during respiration

Results

The live seed in Flask A will have a much higher temperature than the dead ones in Flask B

Conclusions

1. Flask A.

2. To create a semipermeable plug allowing gaseous exchange. Otherwise there is a danger that all the oxygen will be used up, tripping the flasks into anerobic conditions.

3. It gives rise to heat.

4. When taking cuttings the plant seeks to heal itself through producing new roots. This growth is fueled by respiration.

5. Plants grow by using stored energy through respiration. Respiration is a temperature-dependent process; therefore we can raise temperature to the optimum for the crop to fuel our desired growth rates.

6. Photosynthesis uses carbon dioxide and water; respiration uses oxygen and sugar.

Photosynthesis uses energy from sunlight; respiration uses energy from stored starch.

Photosynthesis uses energy to drive chemical reactions; respiration uses energy to maintain and produce new cells.

Photosynthesis occurs only in cells containing chlorophyll; respiration occurs in every cell.

Photosynthesis occurs during the day; respiration occurs day and night.

Photosynthesis occurs in chloroplasts; respiration occurs through mitochondria.

Chapter 7   Plant water

Background

Between 70 and 95 per cent of plant matter is water (by fresh weight). A lack of water is probably the most important factor in the loss of crop yield. Plants use water for a variety of purposes including:

- constituent of cell sap
- participant in a number of chemical reactions
- photosynthesis
- maintains turgidity of cells
- transpiration
- source of protons (hydrogen ions)
- solvent for vital reactions
- medium through which substances move from cell to cell
- medium for the transportation of substances around the plant.

The water content of plants may vary from 70–95 per cent in growing plants, to as low as 5–10 per cent in dormant structures such as seeds. Excluding water, a plant is made up of organic material synthesized from carbon dioxide, water, nitrogen, and the mineral matter absorbed through the roots. Once absorbed, minerals remain in the plant and are only lost through leaf fall. Water, however, is constantly being lost. This affects the rate of absorption and movement of minerals. In order to photosynthesize, plants must get carbon dioxide into the leaf. This is encouraged by transpiration as the plant exposes leaves to the air. This water loss means additional water must be taken up to replace it.

The absorption and transportation of water occurs as the result of several different processes which will be investigated in the following exercises, including:

(a) water movement into plant roots – osmosis, plasmolysis, diffusion.
(b) water movement within the plant – root pressure osmosis, capillary rise.

The following exercises will investigate these processes:

1. model of how osmosis occurs
2. diffusion
3. osmosis
4. plasmolysis
5. osmosis, diffusion and plasmolysis determination
6. root pressure osmosis.

Exercise 7.1

Model of osmosis

Background

Osmosis is the movement of water from a dilute (weak) solution across a semipermeable membrane, to a more concentrated (strong) solution (e.g. the smaller scale movement of water from cell to cell).

A semipermeable membrane (e.g. plant root) is a sieve of tiny pores too small to allow large molecules like sugar ($C_6H_{12}O_6$) to pass through, but large enough to let small water molecules (H_2O) through.

Osmosis is the method by which water is passed into and out of cells. The cells gain or lose water until the solutions are equal. It is therefore important in many functions such as the opening and closing of the stomata and plant turgidity. Air-dried seeds, starches, proteins and cellulose attract water. Water is therefore described as imbibed as water moves from an area of high water content to an area of lower water content. Some people like to think of osmosis as the ability of the plant to suck water into itself.

As water moves across the membrane into the cell, it leads to a build up of pressure, and this develops cell turgidity (turgor). Turgidity enables stems to stand upright and leaves to be held firmly to the light. Normally living cells have a slightly lower than maximum turgor pressure, therefore they can absorb water by osmosis. If water is lost from cells they become flaccid (wilt).

Aim

To predict the movement of solutions of different strengths.

Method

Observe Figures 7.1 and 7.2 and state which way the water will flow (osmotic direction pull). NB: all molecules are in constant motion. The membrane is permeable only to water. The sucrose molecules are too large to pass through.

Results

With reference to Figure 7.1:

1. What will happen to the two solutions?
2. Which way does the water flow?
3. Which parts of the plant might the membrane represent?

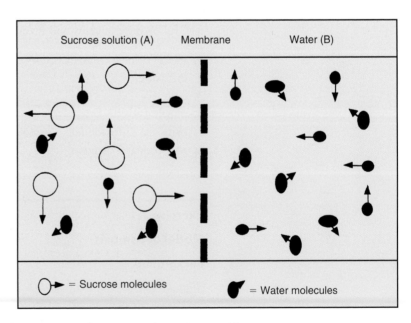

Figure 7.1 Osmotic direction pull for solutions A and B

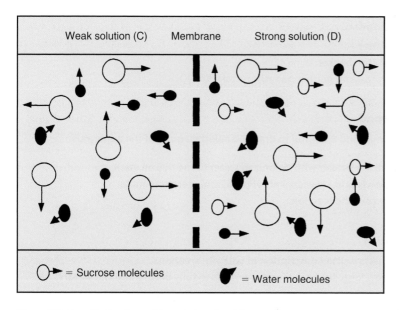

Figure 7.2 Osmotic direction pull for solutions C and D

In Figure 7.2:

4. What will happen to the two solutions?
5. Which way does the water flow?
6. What happens to plant cells when osmosis moves water into them?
7. What would happen if the strong solution was the soil water and the weak solution was the root sap?

Conclusions

1. State, in your own words, what is meant by osmosis.
2. Explain why dried raisins swell when placed in water.
3. Explain how, without drying, you could cause the raisins to shrivel?
4. Explain what would happen to a vase of cut flowers if the owner added more than the recommended dose of Baby Bio plant food to the vase water.
5. How do plants take up water through their roots?

Exercise 7.2

Diffusion

Background

Diffusion is the movement of gas, liquids or salts from regions of high concentrations to low concentrations until the concentrations are equal (e.g. hot and cold water mixed to make warm water).

Diffusion occurs because molecules are in constant motion, trying to produce conditions where they are evenly distributed. It is a slow process, where the rate of diffusion depends on the concentration gradient. This is the difference between high and low concentrations. The greater the difference, the faster will diffusion be. For example, plants transpire water faster through the leaves on a hot sunny day, than on a wet damp day when the diffusion gradient between the moist air outside the leaf and moist air inside the leaf is much narrower. Similarly, if we add fertilizers to the soil, we create a zone of high concentration relative to that surrounding the roots and, once they are in solution (e.g. after rainfall or irrigation), nutrients move towards the roots by diffusion. Some plant processes that utilisze diffusion include:

- water vapor transpired through stomata
- nutrients move through the soil towards roots
- movement of dissolved nutrients between cells
- water movement through the soil to plant roots
- gas movement (CO_2 and O_2) into and out of leaves.

Aim

To demonstrate the process of diffusion.

Apparatus

Water
500 ml beaker
Potassium permanganate crystals (KMNO$_4$)

Method

You may find it helpful to draw a labeled diagram of the apparatus.

1. Fill the beaker with deionized water to the 500 ml mark.
2. Slowly add a few crystals of potassium permanganate.
3. Record the reaction.

Results

1. Give a written description of your observations.
2. How long did it take for the crystals to fully diffuse into the water?

Conclusions

1. Are the following statement true or false?
   (a) 'Water vapor moves by diffusion through the stomata, from an area of high concentration inside the leaf, into the air, an area of low concentration outside the leaf.'
   (b) 'Carbon dioxide moves from the air, an area of high concentration, through the stomata and into the leaf, an area of low concentration, by diffusion.'
2. How does a high atmospheric humidity effect diffusion of water?
3. Explain how the rate of transpiration is effected by:
   (a) spraying pot plants on display
   (b) exposure to wind or draughts
   (c) mist propagation.
4. What is the effect of covering display plants with damp cloth?
5. Write, in your own words, a simple definition of 'diffusion'.
6. Explain how water moves through the soil to plant roots.

Exercise 7.3

Osmosis

Background

Osmosis is the movement of water from an area of low salt concentration, across a semipermeable membrane, into an area of higher salt concentration. It is important in many stages of the water cycle from root to atmosphere including the following.

Entry into the root hairs

Each root hair is a long cell containing a large central area of sap in which various salts and sugars are dissolved. The soil particles are surrounded by a film of water which, although containing salts, is a weak solution compared with that of the cell sap. They are separated by the living cell wall, which acts as a semipermeable membrane. Consequently, osmosis will occur and water enters root hair cells.

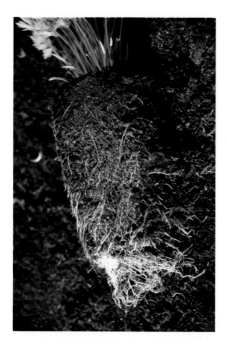

Figure 7.3 Root hairs of carrot seedlings

Passage across the root

The absorption of water into a root hair cell dilutes its contents. There are now two cells of different concentrations next to one another and the root hair cell acts as a relatively weak solution in comparison with the cell internal to it. Water, therefore, passes by osmosis, to the internal cells, which in turn become diluted and so pass water into the cells internal to them. A complete osmotic gradient is established, extending from the root hair cells to the central cells of the root. Water passes along this gradient until it enters the cells around the xylem.

Aim

To demonstrate the process of osmosis.

Useful website

www.biotopics.co.uk/life/osmsis.html

Apparatus

2 potatoes A and B
2 saucers
Copper sulphate crystals
Distilled water

Useful websites

www.biotopics.co.uk

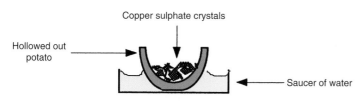

Figure 7.4 Potato experiment

Method

The apparatus should be set out as in Figure 7.4. You may find it helpful to draw a labelled diagram of the apparatus.

1. Boil Potato A for 10 minutes.
2. Hollow out each potato and place on a saucer.
3. Fill each saucer with water.
4. Add a few crystals of copper sulphate to each hollow.
5. Observe the reaction.

Results

Enter your observations in the table provided.

Sample	Hollow description	Saucer description
Potato A		
Potato B		

Conclusions

1. Explain why copper sulphate crystals were used in this exercise.
2. For Potato B, state which area is the zone of high concentration (hollow or saucer).
3. Why is Potato A boiled?
4. Name the process occurring.
5. What would happen if copper sulphate was placed in the saucer and water in the hollows?

6. Describe the process of osmosis and its importance in water movement through plants.
7. How does water move across the plant roots to the xylem vessel?
8. Explain the process involved that enables seeds to absorb water prior to germination?
9. How are stomata guard cells able to open and close?
10. State four functions indicating the importance of water in plants.
11. Explain the relationship between water and plant growth.

Exercise 7.4

Plasmolysis

Background

Plasmolysis is excessive water loss from the plant cell, causing the protoplasm to shrink away from the cell wall.

Figure 7.5 Bitch scorch is caused by high potasium levels in dog urine

Sometimes the soil solution can have a higher nutrient salt content than the root sap and osmosis will occur as water moves out from the plant (low salt level) into the soil (high salt level). The plant cells collapse (become flaccid), resulting in temporary wilting. When placed in a solution of lower concentration (e.g. water), turgor returns as water moves back into the cells. If left too long in the plasmolysis state, permanent wilting occurs and the plant will die. In turf we often see 'bitch scorch' caused by urinating dogs passing high levels of potassium with their water.

Plasmolysis often occurs when seeds are left to germinate in a recently fertilized soil, such as when creating a grass lawn. The seed will dry out, wither and die. If container-grown plants are left to dry out too long between watering, the transpiration from the plant and evaporation from the container will cause an increase in the salt concentration of the solution and plasmolysis may occur. In the alkali flats of the USA, evaporation exceeds rainfall. Salts accumulate in the soil and plasmolysis follows. Water is drawn out from the plants and normal plants cannot grow. Plants especially adapted to survive these conditions are called halophytes (e.g. Yucca). These plants are also adapted to coastal environments, such as the mud flats, salt marsh, and along beaches.

Conductivity

The concentration of salts (nutrients or ions) can be indicated by a single, simple measure of conductivity. Ions that contribute to conductivity readings are nitrate, ammonium, potassium, calcium, magnesium, sulphate, sodium and chloride. High nutrient levels may be detrimental to growth. Plant tolerance to soil salinity varies with species.

Signs of conductivity damage include:

(a) retarded growth
(b) delayed flowering
(c) reduced flowers/fruit size
(d) scorched foliage (overhead irrigation/liquid feed)
(e) wilting.

Some examples are given in Table 7.1.

Table 7.1 Plant tolerance of soil salinity

Very sensitive	Azalea, Camellia, Gardenia, Pittosporum, Primula, Mahonia aquifolium
Sensitive	Aphelandra, Clivia, Erica, Ficus benjamina, Lettuce, most bedding plants, Philodendron, Spartium junceum
Tolerant	Carnation, Chrysanthemum, Cupressus, Diffenbachia, Hydrangea, Magnolia, Hibiscus
Very tolerant	Acacia, Bougainvillea, Atriplex, Callistemon, Cordyline, Dietes, Tomato, Yucca

Aim

To demonstrate the process of plasmolysis in horticultural situations.

Apparatus

Concentrated fertilizer solution (e.g. Tomorite tomato feed or neat Baby Bio)

Deionized water

2 pots of germinating oil seed rape (or other seeds, e.g. cress) seedlings, A and B

Useful website

www.biotopics.co.uk/life/osmsis.html

Method

You may find it helpful to draw a labelled diagram of the apparatus.

1. Drench pot A with deionized water.
2. Drench pot B with concentrated fertilizer solution.
3. Record the reaction over the next hour.

Results

Description of treatment method result

Sample type	Deionized water	Concentrated fertilizer
Rape pot A		
Rape pot B		

Conclusions

1. Name the process that occurred.
2. State what is meant by plasmolysis.
3. What might cause plasmolysis in a horticultural situation? State one method of rectification.
4. Explain why newly transplanted seedlings often wilt.
5. Describe what happens to root cells that are in contact with a soil solution with a high salts level.
6. Explain the impact of dog urine 'bitch scorch' (which is extremely rich in potassium ions) on lawn growth.

Exercise 7.5

Osmosis, diffusion and plasmolysis

Background

Now that you have investigated the processes of osmosis, diffusion and plasmolysis, your skills will be developed to enable you to distinguish each process accurately. In this exercise a semipermeable membrane is used to simulate a plant root placed in different sucrose solutions. A dye is used to aid your observations. Figure 7.6 shows how the experiment finishes when all the reactions have occurred.

Aim

To distinguish between osmosis, diffusion and plasmolysis.

Apparatus

3 × 500 ml beakers
Red dye
Deionized water
3 retort stands and clamps
3 dialysis semipermeable membranes
20 per cent sodium chloride solution (NaCl)

Useful website

www.biotopics.co.uk/life/osmsis.html

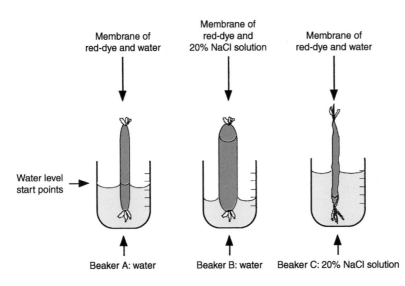

Figure 7.6 Finishing point of experiment showing changes in membrane and beaker water levels from start point

Method

1. Suspend each membrane from the retort stand clamp into a beaker.
2. Prepare the membrane and beaker solutions as follows.
   (a) Beaker A: red dye + water in membrane, water in beaker.
   (b) Beaker B, red dye + 20 per cent NaCl solution in membrane, water in beaker.
   (c) Beaker C, red dye + water in membrane, 20 per cent NaCl in beaker.
3. Record the reaction when the membrane is placed in the beaker, noting, particularly, any changes in water levels.

Results

You may find it useful to draw a diagram showing the apparatus before and after treatment. Enter your results in the table provided:

Situation	Change in water level in beaker	Change in water level in membrane	Interpretation of process occurring
Beaker A			
Beaker B			
Beaker C			

Conclusions

1. State the process occurring in Beaker A.
2. State the process occurring in Beaker B.
3. What would happen to plant cells subjected to the conditions that occurred in the membrane of beaker B?
4. State the process occurring in Beaker C.
5. What would happen to plant cells subjected to the conditions that occurred in the membrane of Beaker C?
6. Describe how water moves:
   (a) through the soil to plant roots
   (b) across the plant roots to the xylem
   (c) Up the plant stem to the leaves.
7. Explain what happens to root cells which are in contact with a soil solution with a high concentration of fertilizer salts.

Exercise 7.6

Root pressure osmosis

Background

Water is aided to move up a plant by root pressure osmosis. This causes water to move through the root and into the xylem vessel of the stem from where it will travel to the rest of the plant. It is induced by the root pumping ions into the xylem.

The cells of the xylem are specialized and empty. They are arranged in columns, one on top of another, so that they form a system of hollow tubes through which water can be passed up the plant. Root pressure osmosis causes water to be passed into the xylem vessel. The water follows passively by osmosis and builds up in the xylem under pressure, normally only two to three atmospheres, occasionally seven or eight bar (one bar approximately equals one atmospheric pressure). Each bar pressure can support about 10 m of water. Root pressure is the main force left to move ions and water up the plant during the night and during the dormant season.

In North America the giant redwoods (*Sequoia*) can be up to 100 m tall. This means that the column of water that must be held in the xylem against the force of gravity is equivalent to the pressure required to lift a weight of 1.6 tonnes resting on the palm of your hand. Part of the force required to sustain such conditions is produced by the osmotic root pressure. This creates a pressure gradient in the xylem and water continues to move upwards through a process called capillary rise.

Figure 7.7 The giant redwoods can grow 100 m tall and carry the equivalent weight of water as 1.6 tonnes resting on your hand

Aim

To demonstrate root pressure osmosis (see Figure 7.8).

Apparatus

Geranium pot plant
Capillary tube
Rubber sleeve
Saucer
Water
Red dye

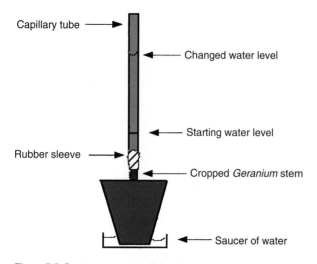

Figure 7.8 Root pressure osmosis in action

Method

1. Crop the geranium at the level of the compost.
2. Attach a capillary tube to the stem via a rubber sleeve.
3. Fill the capillary tube with red dye to 5 cm and draw a line over the meniscus.
4. The plant is watered via the saucer for ten days.
5. Record the change in the capillary water level after 10 days.

Results

You may find it helpful to draw a labelled diagram of the apparatus. Note how many millimetres had the water level has risen in the capillary tubing.

Conclusions

1. Was the results due to:
   (a) photosynthesis
   (b) osmosis
   (c) root pressure osmosis
   (d) diffusion
   (e) plasmolysis?
2. Explain the importance of root pressure osmosis to plant growth and development.

Answers

Exercise 7.1. Model of osmosis
Results

1. Solution A water level will rise; solution B will fall.

2. The water will flow from right to left.

3. Plant root hairs and cell walls.

4. Solution C will fall; solution D will rise.

5. From left to right.

6. They swell and become turgid.

7. Water would move out of the plant into the soil and the plant would wilt.

Conclusions

1. Osmosis is the movement of water from a weak to a strong solution across a semipermeable membrane.

2. Raisins are concentrated sugar enclosed in a semipermeable skin. Consequently they draw water into themselves by osmosis and swell up.

3. The raisins would have to be placed into a more concentrated solution of sugar such as treacle.

4. Plant cells would lose water into the vase as a zone of higher concentration had been established. The plant would wilt and may die.

5. From the soil across the root hair cells' semipermeable membrane and into the cell by osmosis.

Exercise 7.2. Diffusion
Results

1. The crystals diffused from being a solid to slowly dissolved into a liquid, changing the water to a uniform purple colour.

2. One minute.

Conclusions

1. (a) true, (b) true

2. It slows down water loss from the leaf since the diffusion gradient has been narrowed.

3. (a) Slowed to creating a zone of high humidity surrounding the leaves.
   (b) Draughts whip moist air away from the leaf, creating a relatively dry area into which moisture will quickly move from the leaf.
   (c) Creates very high humidity which narrows the diffusion gradient and greatly slows moisture lost from the leaf. This is critically important in allowing the cutting time to grow new roots to absorb water before the aerial part of the plant dries out and dies.

4. It creates high humidity slowing transpiration.

5. Diffusion is the mixing of different concentrations until they are equal.

6. Soil immediately surrounding the root is relatively dry so water moves into this area from a wetter area by diffusion.

Exercise 7.3. Osmosis
Results

Potato A: no change; Potato B: hollow water level rises and saucer level falls as osmosis acts.

Conclusions

1. To create an artificial zone of high concentration.

2. Hollow.

3. As a control to the experiment.

4. Osmosis.

5. Water would move from the hollow across the semipermeable potato into the saucer a zone of higher concentration.

6. Osmosis is responsible for water movement from cell to cell.

7. From cell to cell by osmosis. Each cell in turn becomes diluted relative to the cell internal to it and water flows by osmosis.

8. The seed is a zone of highly concentrated starch reserves. This draws water by osmosis and the seed swells, leading to germination.

9. It is thought that the guard cells regulate the level of ions, especially potassium in them and they swell or shrivel due to the resulting osmosis.

10. Choice from: constituent of cell sap, participant in a number of chemical reactions, photosynthesis, maintains turgidity of cells, transpiration, source of protons (hydrogen ions), solvent for vital reactions, medium through which substances move from cell to cell, or medium for the transportation of substances around the plant.

11. Up to 95 per cent of a plant is water. Most growth. Ninety per cent of plant growth occurs by vacuolation, causing cells to enlarge (see Exercise 5.7).

Exercise 7.4. Plasmolysis

Results

Pot A: no change. Pot B: seedlings rapidly wilted.

Conclusions

1. Plasmolysis sometimes called reverse osmosis.

2. Plasmolysis is excessive water loss from the plant cell, causing the protoplasm to shrink away from the cell wall.

3. Suitable example e.g. over-feeding pot plants rectified by thorough watering.

4. They may have been transplanted into a soil containing a recent fertilizer application.

5. Plasmolysis occurs. Water moves from cell to cell and back out into the soil. The plant wilts and may die.

6. The high potassium ions create a zone of high concentration and water is lost from the grass, causing plasmolysis.

Exercise 7.5. Osmosis, diffusion and plasmolysis

Results

Beaker A: slight colouring of water by dye: diffusion.
Beaker B: water level fell. Tube became turgid: osmosis.
Beaker C: water level rose. Tube became flaccid; plasmolysis.

Conclusions

1. Diffusion.

2. Osmosis.

3. They become turgid.

4. Plasmolysis.

5. They become flaccid.

6. (a) by diffusion, (b) by osmosis, (c) root pressure osmosis, transpiration pull and capillary rise.

7. They lose water by plasmolysis, the plant wilts and may die.

Exercise 7.6. Root pressure osmosis

Results

The water level in the capillary tube will rise many centimeters.

Conclusions

1. (c) root pressure osmosis.

2. It is one of the ways in which plants can help to raise water considerable distances to reach every cell.

Chapter 8 Water transportation pathways and processes

Key facts

1. Water predominately travels through the xylem vessel and from cell to cell by osmosis.
2. Plant stems are specially designed to withstand the weight of water passing through them.
3. Transpiration is water loss through leaf stomata and accounts for over 95 per cent of water loss in plants.
4. Evaporation is water loss from leaf epidermis and accounts for just 5 per cent of water loss in plants.
5. The transpiration stream is the engine driving the movement of water from the roots to the furthest leaf.
6. Transpiration balance is a measure of the difference between how much water a plant absorbs and transpires over the course of a day.

Background

Water in the plant may be considered as one interconnected plumbing network. The lower region is in the soil. Root hairs are in intimate contact with soil water molecules. The upper region is separated from the atmosphere by a thin leaf layer. Plant water is therefore one giant water molecule held together by hydrogen bonds. Therefore loss or absorption at one point effects the whole system.

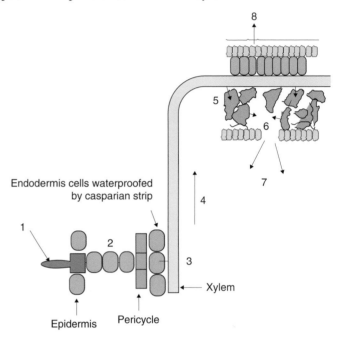

1. Osmosis into root hairs
2. Osmosis from cell to cell
3. Osmosis into the xylem
4. Up the stem by (a) root pressure osmosis
                   (b) capillary rise
                   (c) transpiration stream
5. Into the leaf mesophyll cells by osmosis
6. Diffusion from sxpongy mesophyll into air spaces
7. Transpiration through stomata
8. Limited evaporation through upper epidermis

Figure 8.1 Schematic representation of the transpiration stream

Figure 8.1 shows how the processes concerned with plant water operate in practice to move water through the soil into the plant and out into the atmosphere. This is collectively referred to as the transpiration stream. There are several stages, including:

1. passage through the soil to the plant roots – by diffusion (see Chapter 7)
2. entry into the root hairs – by osmosis (see Chapter 7)
3. passage across the root and into the xylem – by osmosis (see Chapter 7)
4. passage up the stem to the leaves by:
   (a) root pressure osmosis. This enables water to move through the root and into the xylem vessel of the stem from where it will travel to the rest of the plant (see Chapter 7).

(b) capillary rise. As water travels up the plant the xylem vessels narrow, creating further upwards pressure.

(c) transpiration pull/stream. The loss of water by transpiration from the leaves will therefore draw water, by osmosis, from the xylem, which in turn draws further water upwards from the roots.

5. passage out of the plant into the atmosphere by: (a) transpiration and (b) evaporation.

Some of these topics have been covered in the previous chapter. The main focus of this chapter will be on:

- stems as an organ for passage of water up the stem to the leaves
- leaves as organs enabling transpiration into the air
- transpiration in a range of horticultural environments
- methods to reduce water loss from plants.

Exercise 8.1

Stem tissue functions

Background

The stem is the major organ through which water and nutrients are transported around the plant.

Figure 8.2 shows that it contains both unspecialized tissue, such as parenchemya packing/support cells, and specialized tissue, such as vascular

Dicotyledon stem cross section

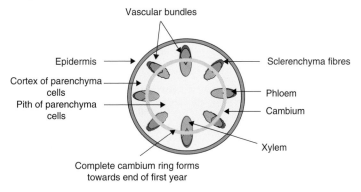

Monocotyledon stem cross section

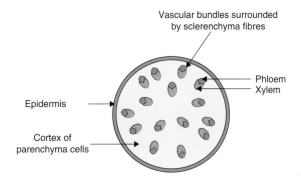

Figure 8.2 Stem primary structures

bundles. Within the vascular bundles of dicotyledonous plants, three types of specialized tissue are found (xylem, phloem, cambium). The xylem vessel is strengthened by a compound called lignin to enable it to withstand the tremendous pressures built up in transporting the weight of water upwards. The phloem is strengthened with cellulose and transports the sugars made during photosynthesis around the plant to where it is needed. The cambium, which is absent in monocotyledonous plants, contains specialized growth cells (meristems) and is responsible for the manufacture of new xylem and phloem cells. In addition, new roots grow from the cambium after cuttings have been taken.

Aim

Investigation into why plants needs stems and the purposes of the tissue that stems contain.

Apparatus

List of tissues including:

Cell membrane	Phloem	Epidermis
Parenchyma	Cytoplasm	Cambium
Vascular bundles	Fibres	Xylem
Endodermis		

Useful websites

www.microscopy-uk.net

www.biologymad.com

Method

Match each stem tissue to the function indicated in the sentences below.

Results

1. Regulates movement of substances in and out of the cell.
2. Protein and water.
3. Vessel for the transport of water up the plant.
4. Vessel for the transport of nutrients around the plant.
5. The growing tips of roots and shoots, small cells producing new xylem and phloem vessels.
6. Cell's protective sheath.
7. Regulates water uptake in the roots.
8. Groups of strands of conducting tissue composed of xylem, cambium, and phloem.
9. Unspecialized soft tissue packing cells.
10. Thread-like cells or filaments.

Conclusions

1. Explain why it is more difficult to take cuttings from monocotyledons than dicotyledons.
2. Describe how xylem tissue is adapted for water transportation around the plant.

Exercise 8.2

Transport through plant stems

Background

One of the plant's four primary organs is the stem. Why do plants need stems and what functions do they perform?

A stem may be defined as the leaf, bud or flower-bearing axis of the plant for support and transport. Sometimes a stem is wrongly called a stalk. A stalk is the common term used for the tissue supporting a flower or leaf. The correct term for a leaf stalk is petiole and for a flower stalk pedicel. A stem can produce buds, a stalk cannot.

Figure 8.3 Stems have many functions

The stem is the main part of the plant above the ground. The functions are:

(a) to hold leaves to the light
(b) to support the flower in suitable positions for pollination
(c) as a pipeline between the roots and leaves, supplying water and nutrients for photosynthesis
(d) a pipeline between the leaves and roots, distributing sugars manufactured by photosynthesis
(e) modified, as organs of food storage and vegetative propagation
(f) in green stems to manufacture food by photosynthesis.

Most stem tips (meristems) grow vertically towards the light (positively phototropic) and away from the force of gravity (negatively geotropic). If the tip is removed (pinched out), growth is prevented (e.g. gladioli to prevent stem twisting). Once it starts growing the plant normally has only one main stem (apical dominance, e.g. tomatoes). This main stem will grow straight upwards rather than in a bushy fashion. In some plants that do not exhibit apical dominance it is necessary to pinch out side buds, preventing shoot growth from these areas and maintaining growth upwards.

Stems have become modified to take advantage of specific environments. Stems of timber, for example, are strengthened to counteract the resistance which the mass of leaves offers to the wind. Stem shapes are also important in supporting the plant such as the square stem of a nettle, or the five-ridged stem of a wallflower. Other plants with naturally contorted stems include twisted willow (*Salix matsudana* 'Tortuosa'), and contorted hazel (*Corylus avellana* 'Contorta'). In dry conditions stems may act as water storage organs (e.g. succulents, cacti and baobab tree).

There are several differences between stems of monocotyledonous and dicotyledonous plants including the content and arrangement of vascular bundles, which help to explain the modifications described above and also influence the passage of water through the plant. This exercise will investigate these differences.

Aim

To investigate differences between monocotyledon and dicotyledon stem structures as the plumbing system of the plant.

Apparatus

Monocotyledon stem slide
Dicotyledon stem slide

Figure 8.4 Twisted hazel

Monocular microscope
Chrysanthemum
Lily

Useful websites

www.olympusmicro.com

www.biologymad.com

Method

1. Cut open stems of chrysanthemum and lily and compare the different tissues seen. These are similar to the tissue which will be enlarged on the slide under the microscope. How does the arrangement of the vascular bundles differ between them?
2. Observe each specimen slide under the microscope/

Results

Draw a fully labelled diagram to show the distribution of tissues in the primary stem of a named dicotyledonous and monocotyledonous plant. Include the following tissues if present:

xylem	vascular bundles
epidermis	cambium
fibres	parenchyma
cell membrane	phloem

Conclusions

1. State two functions of stems.
2. Name two types of cell involved in transporting water and nutrients around the plant.
3. Describe the difference in monocotyledonous and dicotyledonous plant stems with reference to:
   (a) the arrangement of the vascular bundles
   (b) the tissue content of the vascular bundles.
4. State the function of the following parts in the dicotyledonous stem:
   (a) xylem
   (b) cambium.
5. State the functions of the following stem tissues:
   (a) phloem
   (b) epidermis.
6. Where, in a flowering plant, does cell division take place?
7. Where are parenchyma cells found in plants stems?
8. Describe, with the aid of diagrams, the differences between the internal structure of a dicotyledonous and monocotyledonous stem, and relate this to their functions.

Exercise 8.3

Water loss from leaf stomata

Background

Leaves are punctured by air spaces, communicating with the atmosphere by the stomata. The stomata are tiny openings in the epidermis of leaves and allow the entry and exit of gases (mainly carbon dioxide), including moisture vapour, between the leaf and the atmosphere. Through these openings water is able to escape from the plant during transpiration.

Such **gaseous exchange** is essential for photosynthesis and respiration life systems. This occurs through a processes called **diffusion** (see Chapter 7). The cells surrounding air spaces in the spongy mesophyll contain moisture, which will evaporate into the air spaces and diffuse into the atmosphere when the stomata are open.

> The stomata is the main site for **transpiration** without which water and minerals would not be able to move around the plant.

Figure 8.5 Oak leaves have been recorded with over 50,000 stomata per centimetre

Transpiration also aids leaf cooling as the moisture evaporates. For water molecules to change from liquid to gas (vapour) form, they need energy to break their hydrogen bonds. At 20°C this takes 2450 J for every 1 g of water. Consequently, the evaporation of water by transpiration also helps cool the leaf. For example, on a dry, bright summer day, the energy extracted from the leaves of a large rose bush during transpiration is approximately 250 watts.

The opening and closing of the stomata (singular stoma) is controlled by the amount of water in the surrounding guard cells. Increased water pressure makes the cells turgid (bloated). The guard cells maintain turgidity by accumulating ions (especially potassium) in them. Water then moves into them by **osmosis**. Guard cell walls are naturally thickened and the increasing pressure causes the elastic outside walls to stretch and the stomata to open. Sometimes under microscopic examination cellulose micro-fibres can be seen radiating around the circumference. These are called elliptical rings. Generally light causes the stomata to open and darkness causes them to close. Unfortunately, these natural openings into the plant also enable bacterial and fungal spores to penetrate, which may cause subsequent diseases e.g. fireblight (*Erwinia amylovora*). Figure 8.6 depicts a stoma.

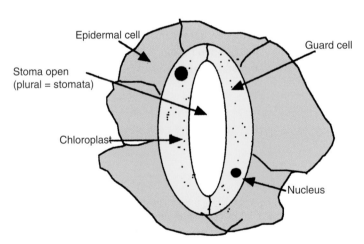

Figure 8.6 Parts of a stoma

Aim

To investigate the variability of stomata between monocotyledon and dicotyledon leaves.

Apparatus

Nail varnish	Slides
Cover slips	Lacto-phenol blue dye
Dicotyledon leaves (e.g. *Hedera* or *Pelargonium*)	Monocular microscope
Monocotyledon leaves (e.g. Grasses, *Tradescanthia* or *Alstroemeria, Narcissus*, most bulbs)	Binocular microscope

Useful websites

www.accessexcellence.org

www.microscopy-uk.net

www.biologymad.com

Method

1. Observe the underside of a monocotyledon and dicotyledon leaf using a binocular microscope. You may be able to see the stomata at high magnification.
2. Paint two thin layers of nail varnish on the underside of each leaf, leaving to dry between coats. When dry, peel off the varnish film and place it on a microscope slide. Add a drop of dye and a cover slip (sometimes results are better without using the dye).
3. Observe the stomata slides under a monocular microscope.

Results

Draw a labelled diagram of your observations including epidermal cell, guard cell, stomata.

Conclusions

1. Estimate, based upon your observations, the number of stomata per square centimetre.
2. Are your specimens stomata open or closed?
3. How did the stomata vary from monocotyledon and dicotyledon specimens?
4. How many elliptical rings could be seen in each guard cell?
5. What is the function of the stomata?
6. For which plant process would the stomata functioning be important?
7. Are the stomata open during the day or at night?
8. How might the plant regulate the opening and closing of the stomata?
9. State the disadvantage of stomatal pores to plant health.

Exercise 8.4

Transpiration

Background

> Transpiration is the engine driving water uptake through the roots of the plant and passing, via the stem, to the leaves where it is lost, by diffusion through the stomata into the air.

Under hot shop lights or windy storage conditions plants may lose more water than they can take up. This results in wilting and the collapse of the leaves and stems as they become limp through the loss of water.

Transpiration is important to plants because it enables:

- water movement around the plant
- nutrients to move around the plant
- leaf cooling.

The majority of water is moved as a result of transpiration. Ninety-five per cent of water loss is through transpiration. Transpiration facilitates the movement of water and nutrients. Nutrients are required at the sites of photosynthesis for which transpiration is essential.

Figure 8.7 Eucalyptus exert a transpiration stream suction of 30–40 bars

The **transpiration stream** describes the movement of water through the plant and its loss through the leaves pulling further water up from the ground. A suction of 30–40 bars (1 bar = 1 atmospheric pressure) has been recorded in *Eucalyptus*, due to transpiration from the leave surface, pulling the water columns up through the xylem. The transpiration stream is the major force causing upward water movement. The column of water in the xylem vessels do not break under this strain, but rather act like a solid, so that the water transmits the pull throughout the whole of its length and the whole column moves upwards. It has been estimated that the suction force set up by the leaf cells as a result of the transpiration stream could raise water to a height of over 300 m.

Water loss through the epidermis of the leaf is called **evaporation.** About 5 per cent of water loss occurs directly from the epidermis cuticle by evaporation. Plants reduce this by producing a waxy coating on the epidermis. These are shiny and help to reflect radiation. Hairs also reduce water loss due to evaporation. Humid air gets trapped between the hairs, lowering the humidity gradient between inside and outside the leaf, slowing down both evaporation and transpiration. Substances called anti-transpirants may be used to spray onto leaves to reduce both transpiration and evaporation. Transpiration rates are influenced by a variety of environmental factors including the following.

Light levels

On bright, sunny days transpiration increases and plants loose more water. This has implications for the amount of irrigation water required to be applied to the plant.

Air temperature

Ambient temperature is the surrounding background temperature.

Transpiration increases with temperature. If we can therefore lower the temperature (e.g. shading or ventilation) we are able to reduce the water loss. At 30°C, transpiration is three times as fast as at 20°C. On average, for every 10°C rise, the rate of transpiration increases 50 per cent.

Air movements

Creating a diffusion gradient between dry and moist air allows diffusion from the leaf (e.g. draughts and central heating systems). Wind-breaks greatly reduce water loss. In windy conditions, the wet air is carried away from the leaf and replaced by drier air and transpiration rates increase. When the air is still, the air surrounding a leaf becomes increasingly wet, thus reducing the rate of transpiration.

Humidity

Air can only take up certain amounts of water before it becomes saturated. As the air becomes wetter (humid), it can take up less water than dry air. If we therefore increase the humidity we can lower the plant's transpiration rate (e.g. misting, spraying or covering plants with a damp sheet). In understocked glasshouses the relative humidity (RH) may drop so low that plants may wilt even if well watered. The optimum relative humidity is 50–80 per cent. RH decreases by 10 per cent for every 1°C rise in temperature.

Soil/container moisture content

Water lost through tranpiration must be replaced by that taken up through the soil. As soil moisture deficit increases, transpiration decreases, and the plant is under increasing stress, losing turgor, and may commence to wilt.

Aim

To assess the transpiration rate of *Garrya elliptica*, under different environmental conditions.

Apparatus

Bubble potometer (see Figure 8.8), comprising:

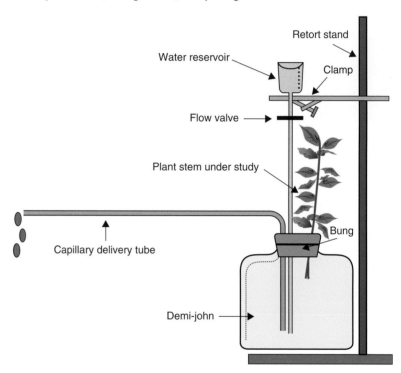

Figure 8.8 Bubble potometer

Retort stand and clamps	Reservoir of water
Capillary tube	Valve
Bung and demi-john	*Garrya elliptica* cutting
Hair drier	Ruler

Useful websites

www.picotech.com

www.saps.plantsci.cam.ac.uk

www.biologymad.com

www.nature.com

Method

1. Set up the apparatus as shown in Figure 8.8.
2. Open the reservoir valve and fill demi-john until water backs down the capillary tube and drips roll off the end, then close the valve.

3. Mark the capillary tube 1 cm from the end.
4. Subject the plant to six different environmental conditions and record how much moisture is used over 5 minute time periods, by recording the distance travelled by the meniscus from the mark made in step 3.
5. It may be necessary to reset the potometer to the end (step 2), if transpiration rates are rapid and the meniscus advances towards the end of the capillary tube.

Results

Enter your results in the table provided. The actual volume of water travelling up the capillary tube may be calculated from the formulae:

$$\pi r^2 h$$

where $\pi = 3.142$, r = radius of capillary tube (1 mm), h = distance travelled by meniscus (mm).

Calculate the actual volume of water passing through the capillary tube over the 5 minute period. Show your working, following the format given in the results table.

Sample conditions	Transpiration loss of moisture over 5 minutes (mm)	Volume of water transpired (ml) $= \pi r^2 h$
Example environment	600 mm	$3.142 \times 0.1 \times 0.1 \times 60 = 1.88$ ml
Room temperature		
High hot air movement		
Outside in shade		
Outside in brightness		
Outside in a cold wind		
Other		

1. Produce a bar chart, on graph paper, to show how transpiration varies with environment.
2. Which environment produced the greatest rate of transpiration?
3. Comment on the significance of your answer to 2 for garden centre retailed plants.
4. Substances called anti-transpirants are available from garden centres. When sprayed over the plants, they produce a waterproof covering for leaves. Explain why anti-transpirants are used in each of the following situations:
   (a) preventing wilting in greenhouse crops
   (b) assisting cuttings to root
   (c) reducing the fall of needles from Christmas trees.

Conclusions

1. Define the term 'transpiration'.
2. Describe four environmental factors that affect the rate of transpiration.
3. Describe what happens to plant cells when plants suffer from a lack of water.
4. State two methods to relieve plants that are wilting.
5. Describe three possible ways in which water may rise from the ground roots to the top of a large tree and mention the relative importance of each mechanism.

6. Warm air currents produced by domestic heating systems have a drying effect on the air. Explain the effect this has on the transpiration rate of house plants.

7. Describe and explain how humidity can be controlled when raising plants.

8. Describe how the transpiration rate may affect the practices of the grower when:
   (a) propagating from cuttings
   (b) growing vegetables in an exposed area
   (c) producing cucumbers under glass.

9. Explain how the rate of transpiration is effected by the following:
   (a) mist propagation environment
   (b) exposure to wind
   (c) damping down
   (d) reduction in temperature
   (e) bright sunshine.

10. State six methods that you could use to reduce water loss from growing plants.

11. Describe how water moves:
    (a) through the soil to plant roots
    (b) across the plant roots to the xylem
    (c) up the plant's stem to the leaves
    (d) Out of the leaves into the atmosphere.

12. Describe, using a diagram, how you would carry out an experiment to test the rate of transpiration from a cutting.

Exercise 8.5

Transpiration balance

Background

Plants may be losing water through transpiration and absorbing it through the roots.

In early morning plants are most turgid and in late afternoon most flaccid. They may be wilting considerably because they have transpired more water than they have absorbed, or they might be experiencing 'guttation' – excess water entering the plant which is removed by 'weeping' leaves. Knowing the transpiration balance greatly aids plant management, e.g. knowing what time of day to take cuttings or harvest plants. Table 8.1 demonstrates this, showing the data obtained from an experiment to compare a plant's rate of transpiration with the rate of water absorption, over a 24 hour period on a hot summer's day.

Aim

To investigate environmental influences on plant water processes.

Apparatus

Graph paper
Ruler
Pencil
Rubber

Table 8.1 Transpiration balance data

Time	Water absorbed in 4 hour period (g)	Water transpired in 4 hour period (g)
04.00	6	1
08.00	6	8
12.00	14	20
16.00	22	20
20.00	13	10
24.00	8	3

Method

On a sheet of graph paper, draw a line-graph to present these results. 'Time', as the constant, should always go on the horizontal *x*-axis. The variable (g of water absorbed/transpired) should go on the *y*-axis.

Results

1. During the day, which rises first, water absorbed or water transpired?
2. What does your answer to 1 suggest about the 'driving force' behind transpiration?
3. Suggest reasons to explain why water continues to be absorbed during darkness/night time.

Conclusions

1. Describe the differences between the amount of water absorbed and transpired at different times, and the effect this may have on the plant.
2. Explain, giving reasons, when plants bought as cut flowers should 'ideally' be harvested.
3. How does this knowledge influence the timing of when to take cuttings for propagation?
4. Distinguish between transpiration, evaporation and evapotranspiration.

Exercise 8.6

The weather

Background

In Britain the weather is a major force shaping the growth and appearance of plants. For example, the shape of growing trees is largely sculptured by wind action. Shading results in the loss of lower branches. They rarely exceed two hundred years of age and are mostly lost through storm damage. Similarly, a common feature of town houses with small fenced garden lawns is algae and fungal disease growth. This is encouraged by the persistent wet conditions and high humidity created by high fences, reducing air movement and slowing evaporation from the soil and transpiration from the grass.

Meteorological data is an extremely useful tool in maintaining the health and establishment of plants. For example, rainfall data is the basis of planning irrigation, and temperature data aids the timing of fertilizer applications. Weather forecasting is supported by information provided from a network of meteorological

Figure 8.9 Details of UK weather can be requested from the Meteorological Office

weather stations spread across Britain, reporting to the central Meteorological Office in Berkshire.

Many controlled environments used in plant production maximize plant response to the processes of photosynthesis, respiration and transpiration. This is through computerized control in the internal climate of the structure, in response to the daily weather conditions.

Aim

To gain an appreciation of the variety of techniques available to record British weather patterns.

Apparatus

A suitable weather station comprising:
Stevenson's screen
Soil thermometers
Anemometer
Grass thermometer
Rain gauge

Useful websites

www.metoffice.gov.uk

www.weatheronline.co.uk

Method

Visit the weather station and record the weather today.

Results

Enter your results in the table provided

Recording	Value
Stevenson's screen:	
Maximum temperature	
Minimum temperature	
Soil temperatures:	
Grass minimum temperature	
Soil at 10 cm	
Soil at 30 cm	
Rain gauge: mm of rainfall	
Anemometer (wind speed)	

Conclusions

1. How does temperature vary between above and below ground?
2. What does this mean in terms of crop growth in spring?
3. What effect will today's weather have on the transpiration rate of plants?

Answers

Exercise 8.1. Stem tissue functions

Results

1. Cell membrane

2. Cytoplasm

3. Xylem

4. Phloem

5. Cambium

6. Epidermis

7. Endodermis

8. Vascular bundles

9. Parenchyma

10. Fibres

Exercise 8.2. Transport through plant stems

Results

Diagrams with labelled tissue similar to Figure 8.2.

Conclusions

1. Choice from: to hold leaves to the light; to support the flower in suitable positions for pollination; as a pipeline between the roots and leaves, supplying water and nutrients for photosynthesis; a pipeline between the leaves and roots, distributing sugars manufactured by photosynthesis; modified, as organs of food storage and vegetative propagation, or in green stems to manufacture food by photosynthesis.

2. Xylem and phloem.

3. (a) Vascular bundles are ordered in dicotyledons and scattered in moncotyledons.
   (b) Vascular bundles are composed of xylem, phloem and cambium tissues except in moncotyledons, which do not have cambium tissue.

Exercise 8.4. Transpiration

1. A suitable bar graph would show typical results in order of greatest transpiration: high hot air movement; outside in bright sunshine, outside in a cold wind, room temperature and outside in shade.

2. High hot air movement.

3. Hot weather and windy conditions will mean that transpiration will be high and additional irrigation of container plants will be required.

Conclusions

1. Monocotyledons do not contain the specialist cambium growth cells which help to produce new roots in dicotyledonous plants when cuttings are taken. It is possible for parenchyma cells to root but requires very careful propagation techniques. More often monocotyledons are propagated by micropropagation.

2. Xylem cells are laid end to end forming a hollow cylinder running through the stem. They are dead cells, hollow and strengthened by lignin, enabling the plant to withstand the weight and pressure of water. This enables water to be transported as a continues film from the root up to the highest leaf.

4. (a) The xylem transports water around the plant.
   (b) The cambium is specialized growth tissue, producing new xylem and phloem cells.

5. (a) The phloem transports sugar made in the leaf down and around the plant.
   (b) The epidermis is for protection and holds everything together.

6. Specialist growth cells called meristems.

7. Packing cells separating vascular bundles and cambium ring from the epidermis.

8. Reproduce diagrams as Figure 8.2. Key points to discuss are that dicotyledons have cambium ring and vascular bundles arranged in a ring. Monocotyledons have no cambium and scattered vascular bundles.

4. (a) Anti-transpirants will slow down transpiration of greenhouse crops.
   (b) Sprayed onto the aerial parts of cuttings anti-transpirants slow down shoot growth. This gives time for new roots to form which can adequately meet the growing plants water requirements.
   (c) Anti-transpirants will prevent the needles from drying out and they will last longer in the home.

Conclusions

1. Transpiration is the loss of water through the stomata in the leaf lower epidermis.

2. Choice from: solar radiation, air temperature, air movement, humidity or soil moisture levels.

3. They become flaccid and the plant may wilt.

4. Overhead irrigation or plunging pot plants in a bucket of water to return the growth medium to field or container capacity.

5. (a) Root pressure osmosis; (b) transpiration pull; (c) capillary rise plus suitable explanations.

6. This will increase the diffusion gradient between the relative wetness inside and outside the leaf and transpiration will increase rapidly.

7. Humidity can be controlled to check growth for example when taking cuttings mist propagation or covering plants with a fleece raises humidity and slows transpiration.

8. (a) Cuttings are very susceptible to wilting and aerial growth must be checked by raising humidity levels, giving new roots a chance to grow and supply the shoots with water.
    (b) Additional irrigation will be needed and a water balance sheet used to calculate irrigation need.
    (c) Glasshouse cucumbers have huge leaves which transpire heavily; ventilation is needed to enable the plants to grow.

9. (a) Misting creates a highly humid environment, slows transpiration and checks aerial growth.
    (b) Exposure to wind widens the diffusion gradient and greatly increases transpiration.
    (c) Damping down lowers the diffusion gradient and slows transpiration.
    (d) Reducing temperature slows transpiration.
    (e) Bright sunshine causes transpiration to increase.

10. (a) By diffusion.
    (b) By osmosis from cell to cell.
    (c) Root pressure osmosis, transpiration pull and capillary rise.
    (d) Transpiration from the lower epidermis and evaporation from the upper epidermis

11. Suitable experiment described similar to Exercise 8.4.

Exercise 8.5. Transpiration balance

Results

1. Transpiration.

2. Daylight and temperature.

3. Cells may be flaccid and water continues to be absorbed by osmosis through the roots to return the plant to a turgid state.

Conclusions

1. Between 06:00 and 19:00 more water is transpired than absorbed. The plant may show symptoms of stress and wilt. Peak transpiration occurs at 16:00. Between 19:00 and 04:00 transpiration is less than water absorbed. The plant will recover the water deficit.

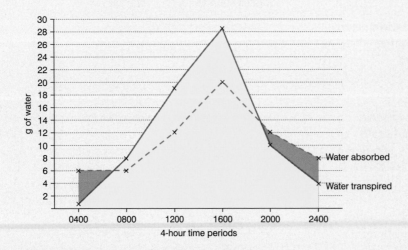

2. They should be harvested early morning since this is when they are most turgid. This will give them maximum shelf life and time for shelf-life conditioning treatments.

3. Cuttings should be taken in the early morning since this is when the plant is most turgid and will give maximum chance of surviving the cutting process.

Exercise 8.6. The weather

Results

Results will vary according to the day and particular weather station visited.

Conclusions

1. Temperature below ground is often warmer than the surface due to microclimate effects. However, the effect may vary seasonally.

4. Transpiration is water loss through the stomata in the lower epidermis. Evaporation is water loss through the cuticle in the upper epidermis and evaporation from the soil surface. Evapotranspiration is the combined water loss from transpiration and evaporation.

2. Plants will start growing in spring when the soil temperature at 10 cm is at least 4°C. This is the temperature at which bacteria and microorganisms begin working at nutrient recycling.

3. Effect needs to be sensibly interpreted in light of particular recordings.

Chapter 9   Flower structure

Key facts

1. Plant sexuality may be either dioecious, monoecious or hermaphrodite.
2. Flower structure diagrams are helpful to describe and catalogue plants.
3. Studies of flower structure can reveal whether plants are dicotyledons or monocotyledons.
4. Flowers may be either insect- or wind-pollinated. There are huge differences in flower structure to reflect this.
5. Pollen is either spread by animals or scattered by wind. Each method is designed perfectly for this adaptation.
6. Fertilization follows pollination and leads to fruiting and seed dispersal.

Background

The blossom is a specialized shoot of a plant bearing the reproductive organs.

Often brightly coloured to attract fertilizing insects, it is the primary organ involved with sexual reproduction. Once the stem apex has begun the shift from vegetative growth to flower development, the change is normally irreversible.

The flower is usually formed from four concentric rings (whorls) – the sepals, petals, male (stamens) and female (carpel) tissue – attached to a swollen base (receptacle) at the end of a flower stalk (pedicel). The female tissues are called gynaecium. The male tissue is called the androecium. Most other flower tissue is concerned with protection and attracting insects.

Figure 9.2 demonstrates a typical half-flower structure, showing the position of all the structures and their role in contributing to sexual reproduction.

Figure 9.1 Brightly coloured blossom of cornflower

Flower sexuality

Flowers contain either male, female or male and female sets of reproductive organs. These are classified as dioecious, monoecious, or bisexual (hermaphrodite) respectively.

Dioecious

Dioecious comes from a Latin word meaning 'two houses'.

These plants are either male or female, i.e. having male and female flowers on separate plants. For fertilization purposes a male plant is usually set among a group of females; both are required if pollination is to take place as self-pollination is impossible. This is not a very common group; examples include asparagus, date palm, *Garrya*, *Ginkgo biloba*, poplar (*Populus*), holly (*Ilex*), Pernettya, Skimmia japonica, willow (*Salix*), yew (*Taxus*), cucumber, hops, spinach, hemp and all cycads.

Monoecious

Having different male and female flowers, but on the same individual plant. This is most common in wind-pollinated plants such as in hazel, where the catkins are the male flowers and the small buds are the female flowers. Self-pollination is unlikely; examples include beech (*Fagus*), birch (*Betula*), hazel (*Corylus*), maize, marrow, oak (*Quercus*), sycamore (*Acer pseudoplatanus*), walnut, and most conifers.

Bisexual or hermaphrodite

Flowers with both male and female organs on the same flower, so do not require a separate pollinator unless the plant is self-sterile. Examples

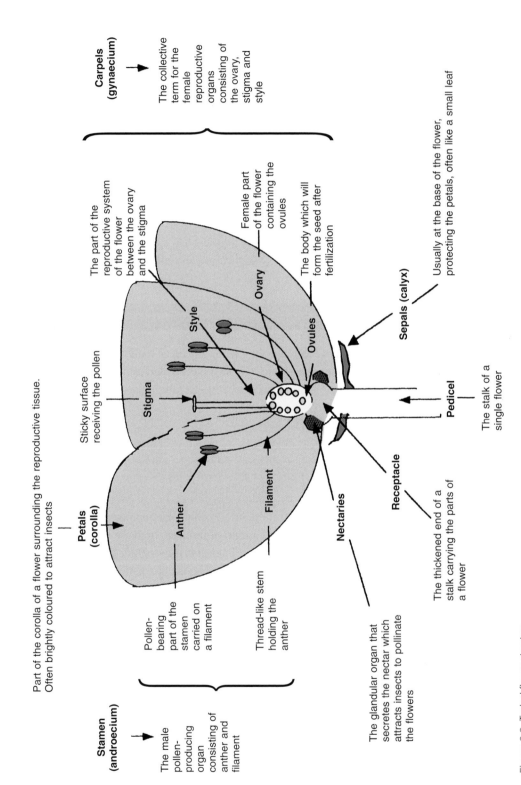

Figure 9.2 Typical flower structure

Table 9.1 Family floral formulae reference list

Aceraceae	K4–5 C4–5 A8 $\bar{G}$ (2)
Aquifoliaceae	K4 C4 A4 $\underline{G}$(4)
Asteraceae	K 0 C(5) A5 $\bar{G}$(2)
Berberidaceae	P3 + 3+3 + 3 A3 + 3 $\underline{G}$1
Betulaceae	P4 or 0 A2–4 $\bar{G}$(2)
Bignoniaceae	K (3–4) C5 A4 G(2)
Boraginaceae	K(5) C(5) A5 $\underline{G}$(2)
Buxaceae	P 4–6 A4 $\underline{G}$(3)
Caprifoliaceae	K(5) C(5) A4–5 $\bar{G}$(2–5)
Carpinaceae	P0 A3–13 $\bar{G}$(2)
Caryophyllaceae	K5 C5 A5–5 G(3–5)
Celastraceae	K4–5 or (4–5) C4–5 A4–5 $\underline{G}$(2–5)
Cornaceae	K4–10 C4–10 A4–20 $\bar{G}$(1–2)
Corylaceae	P0 A4–8 $\bar{G}$(2)
Brassicaceae (Cruciferae)	K2 + 2 C4 A2 + 4 $\underline{G}$(2)
Ericaceae	K4–5 C4–5 or (4–5) A5–10 $\underline{G}$(4–5)
Fagaceae	P4–7 A5–∞ $\bar{G}$(3) or (6)
Geraniaceae	K5 C5 A5 + 5 $\underline{G}$(5)
Hamamelidaceae	K(4–5) C4–5 or 0 A4–5 $\underline{G}$(2)
Hippocastanaceae	K(5) C5 or 4 A8–5 $\underline{G}$(3)
Juglandaceae	P4 A3–40 $\bar{G}$2
Lamiaceae (Labiatae)	K(5) C(5) A4 $\underline{G}$(2)
Lauraceae	(4–6) A8 –∞ $\underline{G}$1
Fabaceae (Leguminosae)	K5 C5 A10 or (10) or (9) +1 or ∞ $\underline{G}$1
Liliaceae	P3 + 3 A3 + 3 $\underline{G}$(3)
Magnoliaceae	K3 C6–9 or ∞ A∞ G∞
Moraceae	P3–4 A3–4 $\underline{G}$1
Oleaceae	K(4–6) C(4–6)
Orchidaceae	P3 + 3 A1 $\bar{G}$(3)
Papaveraceae	K2 C2 + 2 A∞ $\underline{G}$(2–∞)
Platanaceae	P 3–8 A3–8 $\underline{G}$3–8
Primulaceae	K(5) C(5) A5 $\underline{G}$(5)
Ranunculaceae	K5 C5–15 A∞ $\underline{G}$(5)
Rhamnaceae	K4–5 C4–5 or 0 A4–5 $\underline{G}$2–3
Rosaceae	K5 C5 A∞ $\bar{G}$1 or (5) or ∞
Rutaceae	K5–4 C5–4 A10–8 G5
Salicaceae	P0 A2–∞ $\underline{G}$(2)
Saxifragaceae	K5 C5 A5 or 5 + 5 G2
Scrophulariaceae	K(5) C(5) A4 $\underline{G}$(2)
Solanaceae	K(5) C(5) A5 $\underline{G}$(2)
Tiliaceae	K4–5 or (4–5) C4–5 A∞ $\underline{G}$(2–5)
Ulmaceae	P4–5 A4–8 G(2)
Umbelliferae	K5 C5 A5 $\bar{G}$(2)

Figure 9.3 Holly has different male and female plants

Figure 9.4 Hazel: male catkins and female bud

include *Alstroemeria*, buttercup, white deadnettle, bluebell, pea and most plant species.

This chapter will investigate the following areas:

(a) function of the flower
(b) flower sexuality
(c) differences between insect and wind pollinated plants
(d) pollination and fertilization.

Exercise 9.1

Flower dissection

Aim

To gain competence at dissecting flowers and identifying their components tissues.

Apparatus

Fritillaria flowers
Hand lens
Dissecting kit
(suitable alternative flowers may be used e.g. daffodil, lily and *Alstroemeria*)

Useful websites

www.naturegrid.org.uk/qca/flowerparts.html

www.microscopy-uk.org.uk/mag/artnov07macro/flower_anatomy/index.htm

Method

1. Place a fritillaria flower on a white tile and bisect the flower.
2. Draw a diagram of the half flower.
3. Dissect the flower and place each structural part on a piece of card identifying its name.

Results

Draw a labelled diagram of all the tissue present in the half-flower.

Figure 9.5 Tulip has both male and female parts in the same flower

Figure 9.6 Tulip half-flower

Conclusion

1. Define the terms:
   (a) dioecious
   (b) monoecious
   (c) hermaphrodite.
2. Is the plant you have drawn dioecious, monoecious or hermaphrodite?
3. List an example plant that is:
   (a) dioecious
   (b) monoecious
   (c) hermaphrodite.
4. Draw a clearly labelled diagram of the 'typical' half-flower structure of a hermaphrodite (bisexual) flower with which you are familiar.

Exercise 9.2

Structure and function

Aim

To describe correctly the function of flower structure parts.

Apparatus

Previous illustraton of flower parts

List of structure parts:

receptacle	pedicel	sepals
nectar	nectaries	stamens
petals	ovules	ovary
carpels	style	stigma
filaments	anther	

Useful websites

www.floraflora.co.uk/SEFloweranatomy.asp

http://andromeda.cavehill.uwi.edu/flower_structure_and_function.htm

Method

Match the correct flowers tissue with the descriptions of their function.

Results

What is the:

(a) thread-like stem that carries the anther in a stamen?
(b) part of the stamen producing the pollen grains?
(c) female tissue which when receptive is sticky and receives the pollen at pollination?
(d) part of the reproductive system of the flower between the ovary and the stigma?
(e) female part of the flower which will develop into a fruit after fertilization?

(f) female reproductive organs consisting of the ovary, stigma and style?

(g) body which will develop into the seed after fertilization?

(h) part of the corolla of a flower surrounding the reproductive organs, often coloured to attract insects?

(i) male pollen-producing organ consisting of anther and filament?

(j) glandular organ that secretes the nectar?

(k) "beverage of the gods" – the solution secreted by the glands which attracts insects to pollinate the flowers?

(l) tissue, often like a small leaf, at the base of the flower protecting the petals when in bud?

(m) stalk of a single flower?

(n) thickened end of a stalk carrying the parts of a flower?

Conclusions

1. Explain the difference between 'gynaecium' and 'androecium'.
2. Describe the typical arrangement of flower tissue.
3. Describe each of the following structures and state their role in the flower:
    (a) stamens
    (b) stigma, style and ovary
    (c) ovules
    (d) petals.

Exercise 9.3

Floral structures

Background

Flowering plants belong to two main groups, the dicotyledons and monocotyledons. Examination of the flower parts will determine to which group they belong.

Figure 9.7 Hypericum are dicotyledons

Dicotyledons

Flowers that have distinct sepals, petals, stamens and carpels, in multiples of four or five.

Monocotyledons

Flowers that do not have distinct sepals or petals. Instead there are one or two rings of petal-like perianth segments. A perianth is a collective term used to describe the envelope of external floral parts, including the calyx and corolla. The combined petals and sepals of a tulip are called a perianth. The perianth segments of these flowers are in multiples of three.

Ovary – superior or inferior?

Superior: attached to the receptacle above the position of insertion of the corolla whorl and stamens. This flowers structure is called hypogyny e.g. lily.

Inferior: attached to the receptacle below the position of insertion of the corolla whorl and stamens, so that it can be seen below the flower in side view. This flower structure is called epigyny, e.g. daffodils, and *Fuchsia*.

Figure 9.8 Lily are monocotyledons

Flower shapes

Many flowers, if you ignore some finer details, are radially symmetrical in the shape of a starfish. These flowers are called **actinomorphic** or **regular flowers**, e.g. Narcissus, wallflower. There are, however, many flowers that are bilaterally symmetrical with one left hand side and an opposite right hand side. Such flowers are called **zygomorphic** or **irregular** flowers.

Floral formula

A floral formula is a short description of the flower characteristics of the species and often of the family. It is a useful way to store and retrieve plant data. Each flower part is represented by a single capital letter which is followed by a number representing the number of parts present.

K	=	Calyx	(sepals)
C	=	Corolla	(petals)
P	=	Perianth	(fused petals, e.g. tulips)
A	=	Androecium	(male parts – stamen)
G	=	Gynoecium	(female parts – ovary). The ovaries may be superior ($\overline{G}$) or inferior ($\underline{G}$)

When there are a large number of parts (more than 12) the symbol for infinity is used (∞). For example:

$$A\infty = \text{more than 12 stamens}$$

Brackets are used when parts of the flower are joined. For example:

$$K(5) = \text{5 fused sepals}$$

$$C(2) = \text{2 fused petals}$$

If the flower parts are distinctly arranged in more than one whorl, it is shown in a number of ways such as:

$$K\,3+3 = \text{1 ring of 3 sepals} + \text{another ring of 3 sepals}$$

$$A4+2 = \text{4 stamens} + \text{another 2 stamens}$$

When different flower parts are joined such as stamens fixed to the petals, it is show by a line in the formula, for example:

$$C(5) - A4 = \text{4 stamens joined with 5 fused petals}$$

Aim

To compare and contrast the flower of a dicotyledon with that of a monocotyledon.

Apparatus

Poppy	Buttercup
Daisy	Tulip
Daffodil	Bluebell
Alstroemeria	Lily
Dissection kit	Hand lens

Method

1. Dissect and examine each flower.
2. Record your observation in the table provided.
3. Translate the following formulae:
    (a) K2 + 2 C4 A2 + 4 $\bar{G}$ (2)
    (b) K(5) C (5) A (9 + 1) $\bar{G}$1
    (c) K0 C0 A3 G0
    (d) K0 C0 A0 $\bar{G}$(2)

Results

Enter your results in the table provided.

	Poppy	Buttercup	Daisy	Tulip	Daffodil	Bluebell	Alstroemeria	Lily
Petal segments joined or separate								
Number of petal segments								
Number of stamens								
Superior or inferior ovary								
Monocotyledon or dicotyledon								

Conclusion

Describe the floral differences between monocotyledonous and dicotyledonous plants.

Exercise 9.4

Pollen investigation

Background

Pollination is defined as the transfer of pollen (the male sex cell) from the anthers to the female stigma.

When ripe, the pollen sacs of the anther split open and expose the pollen, which can then be dislodged. The pollen is either carried by insects, or blown by the wind to the stigma of another flower. It leads to fertilization and fruit development, resulting in new seed. The practice of dehiscing flowers (breaking open the anthers to release the pollen) is a widespread practice in horticulture. Similarly, the use of an electric bee, a vibrating probe shaking the flower to dislodge the pollen, encourages pollination in glasshouse crops. Many growers hire mobile bumble bee hives to keep in the glasshouse. If pollination is necessary (e.g. in apples) it is important not to apply insecticides or fungicides at full flower.

Generally, plants are either adapted to enable insect or wind pollination and this is reflected in the characteristics of the flower and pollen grain produced. The structure and physiology of a flower are such that the chances of successful pollination by insects or wind are maximized. These adaptations include the following.

Insect-pollinated plants adaptations

The sequence involves an insect visiting one flower, becoming dusted with pollen from the ripe stamen and visiting another flower where the pollen on its body adheres to the stigma.

General flower adaptations	Pollen characteristics
• brightly coloured petals to attract a variety of insects	• small anthers firmly attached to flower
• scent	• anthers lie within flower
• honey guides (dark lines) direct insects to nectar source	• anthers positioned for insect collision
• honey guides force contact with stamens and stigmas	• large conspicuous flowers
• nectar	• smaller quantities of pollen produced wind
	• large pollen grain size
	• sticky pollen
	• spiked/barbed pollen enabling them to stick in clumps on insects

Figure 9.9 Pampus grass seed head

Wind-pollinated plant adaptions

Grasses, for example, are pollinated by air currents. First, the feathery stigmas protrude from the flower, and the pollen grains floating in the air are trapped by them. Subsequently, the anthers hang outside the flowers, the pollen sac splits, and the wind blows the pollen away. The sequence varies with species.

General flower adaptations	Pollen characteristics
• small inconspicuous flowers	• anthers large, loosely attached to filament
• petals small – plain, often green	• stamens hang outside flowers, exposed to the wind
• no scent	• anthers and stigma hang outside the flowers so the slightest air movement shakes them
• no nectar	• large quantities of smooth, lightweight pollen are needed because it is less reliable
• whole flower loosely attached	• smooth, aerodynamic grains
• feathery stigmas act as a net which traps passing pollen grains	

Aim

To investigate the differences between wind- and insect-carried pollen.

Apparatus

Pollen from grass (or other wind-pollinated plants, e.g. pine)
Pollen from fritillaria (or other insect-pollinated plants, e.g. lily, iris, daffodil)
Binocular microscope

Figure 9.10 Grass has small, inconspicuous flowers

Useful website

www.pollenuk.co.uk/aero/pm/WIP.htm

Method

Observe the pollen under the microscope and compare and contrast their structures.

Results

Describe the differences in the two types of pollen. Draw a diagram of your observations.

Conclusions

1. State which flower's pollen grain was the largest in size and explain the significance of this.
2. Describe and comment on the difference in shape of the two types of pollen.
3. Explain why the fritillaria pollen was sticky and grouped in clumps.
4. Describe four adaptions which exist in wind-pollinated flowers.
5. State what is meant by the term 'pollination' and explain its importance to flowering plants.
6. Name two examples of flowers that are pollinated by wind.
7. List three ways by which some plants are adapted to prevent self-pollination.
8. List four structures which exist in flowers and describe their function, if any, in insect-pollinated plants.

Figure 9.11 Insect often pollinate flowers

Exercise 9.5

Insect-pollinated plants

Background

Bees pollinate flowers, allowing fertilization to take place. They feed on the nectar and collect pollen from the stamens. Normally a group of insects will work a group of flowers methodically to collect as much pollen as possible.

Aim

To observe the variety and frequency of insect pollination of flowers.

Apparatus

    Graph paper
    Stop watch

Method

1. Select a small patch of named flowers (e.g. heathers) for observation.
2. Observe all the flowers in the patch and over 15 minutes, and record:
   - (a) the number of bees visiting the patch
   - (b) the time spent by each bee at the patch
   - (c) the number of flowers visited by the bee during this time.

Results

Enter your results in the table provided.

Observations	Patch 1	Patch 2	Patch 3
Number of bees visiting patch			
Average time spent by each bee at the patch			
Average number of flowers visited by each bee			

Produce a bar graph summarizing your data.

Conclusions

1. What was the average result obtained?
2. How might a strong wind have affected your results?
3. What other insects were involved?
4. Name a flower that is pollinated by a large insect.
5. Describe four adaptations which exist in insect-pollinated flowers.

Exercise 9.6

Fertilization

Background

Fertilization in plants follows pollination

The fusion of the male pollen nucleus with the female ovule nucleus to form a seed is called fertilization.

The pollen grain absorbs sugar secreted by the stigma. A pollen tube is formed which grows down through the style. On reaching the ovary, the tube grows to one of the ovules and enters through a hole, called the micropyle. The tip of the pollen tube breaks open in the ovule, and the nucleus, which passes down the tube, enters the ovule and fuses with the female nucleus. Each egg cell in the ovule can be fertilized by a male nucleus from a separate pollen grain. A fertilized ovary is called a **fruit**.

Sometimes fertilization can occur without pollination. This creates parthenocarpic fruit. For example, seedless grapes and bananas are fruit set without pollination occurring. They are chemically treated with a hormone to induce fruit set. Flowers which suffer frost damage (e.g. pear crops), will have their styles damaged, which will prevent the growth of the pollen tube and on inspection the style will be black; fertilization cannot take place. But when sprayed with Berelex (gibberellin hormone) they form parthenocarpic fruit. The fruit will be misshapen, but flavour is not affected.

After fertilization, most of the flower parts (petals, stamen, style and stigma) wither away. Sometimes traces of them may remain (e.g. the receptacle swells and becomes succulent as in apples).

In all plants the main change after fertilization involves the ovary and its contents. Generally, the ovary appears greatly swollen. Inside the ovule, cell division and growth produces a seed containing a potential plant or embryo and the outer ovule wall forms the seed coat (testa). Water is withdrawn from the seeds, making them dry and hard and enabling them to withstand unfavourable conditions. If conditions are suitable the seed will germinate; if not they will remain dormant until the conditions are right.

Figure 9.12 Daffodil, unfertilized ovary

Figure 9.13 Daffodil, fertilized with swollen ovary

Aim

To investigate fertilized and unfertilized flowers.

Apparatus

Fertilized daffodil A and unfertilized daffodil B (or any two flowers)
Microscope
Dissection kit

Useful website

http://theseedsite.co.uk/develop.html

Method

1. Examine the two daffodils on display.
2. Dissect both ovaries.

Results

1. Describe the differences between the two daffodils.
2. Draw a diagram of each ovary, showing how they differ.

Conclusions

1. Explain which plant has been fertilized, giving reasons for your answer.
2. Define the term 'fertilization'.
3. Explain how 'pollination' differs from 'fertilization'.
4. Describe the changes that occur in flowers after fertilization.
5. Describe the methods by which poor fertilization may be improved in horticulture.
6. Briefly describe the sequence of events leading to fertilization and fruit and seed formation.

Answers

Exercise 9.1.   Flower dissection

Results

Diagram similar to Figure 9.2, labelling all the tissues.

Conclusions

1.   (a) Separate male and female flowers on separate male and female plants.
     (b) Separate male and female flowers, but on the same individual plant.
     (c) Flowers with both male and female organs in the same flower.

2.   *Fritillaria* are hermaphrodite

3.   (a) Suitable example, e.g. asparagus, date palm, *Garrya*, *Ginkgo biloba*, poplar (*Populus*), holly (*Ilex*), Pernettya, Skimmia japonica, willow (*Salix*), yew (*Taxus*), cucumber.
     (b) Suitable example, e.g. beech (*Fagus*), birch (*Betula*), cucumbers, hazel (*Corylus*), maize, marrow, oak (*Quercus*), sycamore (*Acer pseudoplatanus*), walnut, most conifers.
     (c) Suitable example, e.g. A*lstroemeria*, daffodil, tulip, buttercup, white deadnettle, bluebell, pea, *Fritillaria*.

4.   Diagram similar to Figure 9.2.

Exercise 9.2.   Structure and function

Results

(a)   Filament

(b)   Anther

(c)   Stigma

(d)   Style

(e)   Ovary

(f)   Carpels

(g)   Ovules

(h)   Petals

(i)   Stamens

(j)   Nectaries

(k)   Nectar

(l)   Sepals

(m)   Pedicel

(n)   Receptacle.

2.   The flower is usually formed from four concentric rings (whorls), the sepals, petals, androecium and gynaecium, attached to a swollen base (receptacle) at the end of a flower stalk (pedicel).

3.   (a) Stamens are the male pollen-producing organ, consisting of anther and filament.
     (b) Female reproductive organs called carpels. The stigma is sticky and receives the pollen at pollination. The style is part of the reproductive system of the flower between the ovary and the stigma. Pollen tubes grow down through the style to fertilize ovules in the ovary. The ovary contains ovules which will develop into seeds when fertilized. A fertilized ovary is called a fruit.
     (c) The ovules are contained in the ovary; when fertilized they will develop into a seed.
     (d) The petals are part of the corolla of a flower surrounding the reproductive organs, often brightly coloured to attract pollinating insects.

Conclusions

1.   Androecium is the collective term for the male reproductive tissues. Gynaecium is the collective term for the female reproductive tissues.

Exercise 9.3.   Floral structures

Results

The dicotyledons are buttercup, daisy and poppy. The monocotyledons are *Alstroemeria*, bluebell, daffodil, lily and tulip.

(a)   1 ring of 2 sepals + another ring of 2 sepals. 4 petals. 1 ring of 2 stamen + another ring of stamens. 2 fused carpels. Superior ovary.

(b)   5 fused sepals, 5 fused petals, 1 ring of 9 fused stamen + another fused stamen, superior ovary.

(c)   No sepals, no petals, one ring of 3 stamen.

(d)   No sepals, no petals, no stamen, one ring of 2 superior ovary stigmas.

Conclusions

Monocoyledon flowers have their parts in multiples of threes, dicotyledons in multipes of fours or fives.

Exercise 9.4.   Pollen investigation

Results

Diagrams showing small, light, aerodynamic grains of wind-pollinated-plant's pollen and large irregular, clumpy pollen of insect-pollinated-plant's pollen.

Conclusions

1.   Insect-pollinated plant pollen was the largest, enabling them to easily stick to insect.

2.   Insect-pollinated plant pollen is irregular in shape, often containing barbs. These help the grains to stick to insects and increase the chances of successful pollination. The wind-pollinated plant pollen are round, tiny and often contain air sacs. These enable them to be easily blown around in the wind.

3.   Sticky pollen is a characteristic of insect-pollinated plants. These ensure maximum opportunity to attach themselves to insects and so increase the likelihood of being transported to another flower where pollination might take place.

4.   Any four from: small inconspicuous flowers; anthers large, loosely attached to filament; petals small – plain, often green; stamens hang outside flowers exposed to the wind; no scent, no nectar; anthers and stigma hang outside the flowers the slightest air movement shakes them; whole flower loosely attached; feathery stigmas acting as a net which traps passing pollen grains.

5.   Pollination is the transfer of pollen from the anther to the stigma. It is followed by fertilization and fruiting which leads to new seed production. Through pollination cross-fertilization can occur, which leads to improved plant characteristics such as resistance to disease.

6.   Suitable examples such as birch and grass.

7.   Self-pollination may be prevented by pollen being self-sterile, inhibitors to pollen tube growth present in the style, and by differential timing of the ripeness of the stigma and the pollen.

8.   Any four structures from Figure 9.1 listed and described.

Exercise 9.5.   Insect-pollinated plants

Results

These will vary with time of year and particular patch of plants observed.

1.   To calculate the average use the formulae: (total number of bees visiting)/3.

2.   Strong winds typically reduce the numbers of bees visiting.

3.   Typically hoverflies, ladybirds and wasps might be seen.

4.   Any suitable example, e.g. orchid pollinated by a bee.

5.   Any four from: brightly coloured petals to attract a variety of insects; small anthers firmly attached to flower; scent; anthers lie within flower; honey guides (dark lines) direct insects to nectar source; anthers positioned for insect collision; large conspicuous flowers; and honey guides force contact with stamens.

Exercise 9.6. Fertilization

Results

The fertilized ovary will be swollen and individual swollen ovules can be seen. The flower parts have mostly withered or dropped off completely.

The unfertilized daffodil will still have visible petals and floral parts. The ovary will be slim and the ovules poorly differentiated.

Conclusions

1. The fertilized ovary is visibly fatter and swollen.

2. Fertilization is the fusion of the pollen grain nucleus with the ovule nucleus in the ovary.

3. Pollination is the transfer of pollen from the anther to the stigma. The pollen grain then germinates and grows down through the style into the ovary where the nucleus fuses with an ovule nucleus. Only then will fertilization have taken place.

4. After fertilization the ovary swells up as the seed develops. Most of the floral parts shrivel and drop off as the purpose in attracting pollinating insects has been fulfilled. Sometimes flower tissue may swell up e.g. the receptacle in an apple, to encircle the fruit and aid dispersal.

5. Poor fertilization may be overcome by spraying hormones to induce fruit set.

6. Pollination occurs when the pollen is transferred from the anther to the stigma. After the pollen grain has grown down through the style and fused with the ovule, nucleus fertilization has occurred. A fertilized ovary is called a fruit and contains the seed which will be dispersed by wind or animals.

Part Two

Soil science

Chapter 10   Soil formation and texture

Key facts

1. Soil forms from metamorphic, igneous and sedimentary rock.
2. There are seven soil formation factors including parent material, climate, topography, organisms, time and man.
3. Soil formation stages include weathering, colonization, litter-layer, soil depth, organic matter, stabilized soil.
4. Texture is the single most important soil property and is the relative percentage of sand, silt and clay particles present.
5. Clay particles have a surface area 1000 times greater than that of sand.
6. Soil texture will influence how we manage the soil and what plants can be grown.

Background

Soil, like all growth media, enables plant roots to grow. A good soil provides adequate oxygen, nutrients, water, and anchorage. Soil management is dependent upon the varying proportions of mineral matter, organic matter, water and living organisms present.

Soil formation

The type of soil developing in any given place is determined through the interaction of six major factors: parent material, climate, topography and relief, organisms, time and man.

There are three types of rocks from which soils forms: igneous, sedimentary and metamorphic.

Parent material

All mineral matter comes from the weathering of rocks producing soil. Parent material also influences the rate at which soil-forming processes take place. The rocks will be broken down, moved, sorted and deposited, influencing the final soil characteristics found including soil structure, texture, porosity, minerals present and water-holding capacity.

Igneous rocks (e.g. granite) are slow to erode, often forming infertile sandy or loamy soils. Basalt is rich in minerals and tends to form clay or loamy soils which are more fertile.

Sedimentary rocks (e.g. sandstones and chalk) form sandy soils which are often deep, potentially fertile and easily cultivated. Siltstone deposits settle out of water. They feel smooth and form silty soils which, although they have a high water content, are structurally unstable if not carefully cultivated. Limestone contains particles or microscopic shells. Chalk is a special form of limestone, which is very easily weathered by rain to form shallow dry limey soils.

Metamorphic rocks have undergone chemical changes over time (e.g. granite changing to schist; shales changing to slate). Slate, as an example, tends to form infertile silty clays.

Climate

Rocks are the origin of most of the mineral matter in the soil. Weathering is the breakdown of rocks and rock minerals which may subsequently move by erosion. Climatic factors include temperature, rainfall (precipitation), wind and frost action. These interact to determine the nature of weathering chemical and physical processes.

The rate of weathering is temperature-dependent. Moisture must penetrate the soil to act on the parent material. If too warm, moisture may evaporate from the soil surface and

Figure 10.1 Profile showing soil and parent material

there will be little soil development. Cool temperatures with frequent rainfall better allow moisture to penetrate the soil and development advances. An extreme range of temperatures between sunny days and very cold nights causes stresses in the rock which in time result in surface layers peeling away. The continued expanding and contracting of rock surfaces has a weakening effect on the rock.

Rainwater contains carbonic acid which has a very slow but powerful effect dissolving rocks into minerals. Water also expands on freezing. Any water present in cracks in the rocks will freeze and expand to open up the rock.

Heavy rainfall may result in leaching and the formation of pans. Where evaporation exceeds rainfall a moisture deficit exists. This results in upward (capillary) movement of water and salts through the soil and the saline conditions may restrict crop growth.

Moving water (e.g. streams, rivers, waves on the sea shore) all have an abrasive (grinding) effect. Any rock fragments in the water are exposed to the moving water and will be eroded and ground down even smaller. Eventually, coarse sharp fragments are worn down to smooth stones and fine sand and silt.

Organisms

All organisms, such as bacteria, fungi, vegetation and animals, change the chemical and physical environment of the soil. Vegetation protects the soil from wind and water erosion. Plant roots penetrate downwards through the parent material which, together with microorganism activity, split parent material apart. Roots also secrete weak organic acids (e.g tannic, acetic, humic) which attack the rocks and start to dissolve away their softer parts.

Mosses and lichen acidify rock material, enabling weathering to take place. Leaf mineral content provides organic mater which improves structure. Hardwoods absorb bases (calcium, magnesium,

Figure 10.2 Lichen colonizing a rock

and potassium) and recycle them back to the soil surface during leaf fall. Conifers are low in bases and create more acid soils. Some microorganisms promote acid conditions. Earthworms and burrowing animals mix and aerate soil.

Topography and relief

The shape of the land, slope and aspect creates difference in relief which influences drainage patterns through affecting run-off and depth to water table. Higher elevations and sloping areas create well or excessively drained soils. At lower elevations soils receive surface runoff from higher elevations and often have a seasonal high water table. Poorly drained areas often show poor structure with signs of mottling and gleying present.

Time

The length of time taken to form soil is a matter of conjecture. However, the process is continuing and destroyed soil is not easily restored.

Anthropogenic factors (man)

Man's activities have greatly altered soils. Ploughing, cultivation, the addition of lime and fertilizer have changed the physical and chemical properties of soil. Good management (e.g. drainage systems and tillage works) can maintain or improve the soil. However, often poor management techniques have resulted in erosion and the loss of productive land. Examples include the dustbowls created in the USA during the 1930s and the reduction in North American topsoil from 4 metres to 5centimetres in some places, since early settlers started farming.

Stages of soil formation

First weathering produces thin layers of rock fragment which are then colonized by simple 'pioneer' plants such as lichens and algae. A litter layer forms from decaying plant tissue which becomes the food of other organisms (e.g. fungi and bacteria).

The organic matter of the litter layer then becomes a source of nutrients for other plants. New root growth prevents particles from blowing away to leave exposed bare rock. Soil depth increases, enabling higher plants, which have deeper rooting requirements, to colonize. This causes further disintegration of rock at greater depth caused by the larger roots which can explore deeper down into the underlying rock. Consequently organic matter and organisms such as earthworms occur at increased depth, creating stabilized mature soil.

Soil texture

Soil texture is the single most important soil property. Most aspects of soil management are related to it. Texture is the relative proportions by weight of sand, silt, and clay in the soil. This is a basic inherited property of a soil that cannot be changed easily by cultivation.

The mineral component of soils can be split into three types (sand, silt, and clay), defined solely by size. There are several different classifications but the one now being used by ADAS (Agricultural Development and Advisory Service, a division of the Ministry of Agriculture Fisheries and Food (MAFF)), is as in Figure 10.3 and Table 10.1.

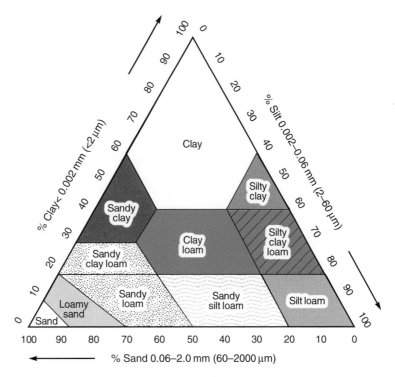

Figure 10.3 Proportions of sand, silt and clay in soil.

Table 10.1 Assessing particle size

Particle	Symbol	Subdivision and textural feel	Particle size (diameter), mm
Sand	S	Coarse: easily seen with the naked eye. Feels and sounds very gritty	0.6–2.0
		Medium	0.2–0.6
		Fine: Only slightly gritty. Difficult to detect in soils and easily missed	0.06–0.2
Silt	Z	Silky, smooth, often buttery when wet	0.002–0.06
Clay	C	Stiff to the fingers; very sticky when wet	0–0.002

Technically, anything larger than 2 mm becomes a stone and is not 'soil'. In addition if a soil contains more than 45 per cent organic (peaty) matter it is classified as an organic soil rather than a mineral soil and different methods to determine texture must be employed.

The symbol 'L' may be used to represent a loamy soil (e.g. ZCL – silty clay loam). Soil texture may be determined by working the soil between the fingers. Each textual class can be distinguished by its moulding properties. There are a range of sizes present within these classes and texture assessment helps us to understand the working properties of our soil including:

(a) Is the soil easy to dig? (structure)
(b) Will it grow plants easily? (cultivations)
(c) Will it hold water easily?
(d) Will it be 'fast' or 'late' to warm up in the spring?
(e) Will it drain well?
(f) Application rates of pesticide and herbicides.

These exercises are designed to introduce the technique of hand texture assessment using the ADAS system modified in conjunction with the Soil Survey of England and Wales in 1986. This is the method used extensively in land management disciplines in the UK.

Figure 10.4 Particle size

Exercise 10.1

Properties of soil particles

Background

Soil texture is the relative proportions of sand silt and clay in the soil. The particles of these elements vary in size (see Figure 10.4).

Sandy soils (0.06–2.0 mm)

These soils feel gritty. They are easy to dig but it is often difficult to grow 'good' plants in them. They are well drained and aerated. These soils do not hold plant nutrients at all well. They have a low water-holding capacity and water drains through the soil quickly.

Sandy soils are prone to drought and in dry weather they dry out quickly. Plants find it difficult to get enough water. It is important to remember that the limits of sand particles covers a wide range of diameters. Coarse sand will drain much faster than medium of fine sand. This is important when choosing sands for improved drainage or for formulating loamless composts mixes. Sandy soils are good for draught-tolerant plants e.g. bearded iris, verbascum species, onion, carrots, poppy, geranium species, convallaria majalis, lupins, potentilla, sea pink, sedum species.

Silty soil (0.002–0.06 mm)

These are stone-free, deep soils. They feel silky and soapy. They are not very sticky and not very gritty. They can be difficult to cultivate and may cause drains to block as they 'cake' the porous drains with layers of sediment.

Their workability is often improved by adding bulky organic matter, compost or farmyard manure. Normally they hold enough water and nutrients to sustain plant growth.

Figure 10.5 Poppy: an ideal plant for sandy soils

Figure 10.6 Pyracantha are able tolerate clay soils well

Clay soils (less than 0.002 mm)

These feel sticky and buttery when wet. They are very difficult to dig when dry. The best time to dig them is when they are moist, but not too wet. Clay particles have good colloidal (chemical) properties and are able to hold on to large quantities of nutrients, making them very useful for growing good plants. The particles are small but this gives them about 1000 times the surface area of the larger sandy particles. They are often called **'cold'** or **'late'** soils because the water which lies in the spaces between the soil particles (pores) warms up slowly in spring. As they absorb water they become very sticky and unworkable. When they dry out, they shrink, becoming hard and cracked. Plants suitable for growing in clay soils include: cotoneaster , dogwood, gunnera hosta, helianthus, pyracantha and willow.

Loams

These soils provide a mixture of soil particles and are considered by many people as best soils to grow plants in. They contain roughly equal proportions of sand, silt and clay. They are just about ideal growing environments and most plants will grow well on them.

Aim

To distinguish by feel the relative properties of sand, silt and clay.

Apparatus

> sandy soil
> silty soil
> clayey soil
> water bottle

Useful websites

www.physicalgeography.net

www.rhs.org.uk

Method

1. Place a spoonful of soil in the palm of your hand.
2. Rub the soil together between finger and thumb.
3. Record what the soil feels like.
4. Think of a similar material which feels the same as the soil. This helps you to remember the characteristic.

Results

Record your results in the table provided. Suggested comparable materials include salt, pepper, talcum powder, Plasticine, soap, butter and others.

Soil type	Texture (feel) characteristics	Material with a similar feel
Sand		
Silt		
Clay		

Conclusions

1. State the accepted size range for the classification of soil particles into:
   (a) coarse sand
   (b) medium sand
   (c) fine sand
   (d) silt
   (e) clay.
2. Describe the properties of each of the following;
   (a) sand
   (b) silt
   (c) clay.
3. Select the appropriate word to describe how clay particles can be recognized when handled between the fingers and thumb:
   (a) crumbly
   (b) gritty
   (c) silky
   (d) sticky.
4. State two benefits of clay as an ingredient in soil.
5. Describe how sand may be identified by feel and state one advantage and one disadvantage of it as a constituent in soil.

Exercise 10.2

Features of coarse, medium and fine grade sand

Background

Sandy soils are mostly quartz and silica mineral grains resistant to weathering. They contain large air spaces and have poor water-holding capacities. For this reason, medium sand in particular is useful as a top dressing on a newly spiked lawn to increase aeration. Because of the large diameter size of coarse sand, it may sometimes be useful to mix with a clay soil and increase the number of larger pores, aiding drainage and aeration. They are chemically inactive (inert), making them sterile. Because of the lack of excess moisture and large air spaces, they warm up quickly in the spring but may be prone to drought in summer.

Sands fall into three size ranges:

Coarse: 0.6–2.0 mm (600–2000 μm)
Medium: 0.2–0.6 mm (200–600 μm)
Fine: 0.06–0.2 mm(60–200 μm)

Well-drained sandy soils produce the ideal conditions for the breakdown of organic matter by bacteria, and are often referred to as '**hungry**' soils because they require a great deal of supplementary nutrient feeding (fertilizers) and mulching.

Aim

To distinguish between coarse, medium and fine grade sand.

Apparatus

Binocular microscope
Glue
Coarse, medium and fine grades of sieved sand

Method

1. Rub each sand grade together between finger and thumb.
2. Record what the sand grade sounds and feels like.
3. Think of a similar material which feels the same as the grade. This may help you to remember their characteristics.
4. Observe and describe each grade under the binocular microscope.
5. Glue a small sample on to the sheet for future reference.

Results

Record your results in the table provided.

Sand grade	Texture (feel) characteristics	Audible (sound) characteristics	Microscope observations	Sample specimen
Coarse				
Medium				
Fine				

Conclusions

1. Which sand would you use as a top dressing on an ornamental lawn?
2. Which sand would you use to improve the properties of a clayey soil?
3. What is the effect of adding organic matter on sandy soils?
4. Which of the following is essential to allow water to drain through soil:
   - (a) a continuous film of water round soil particles
   - (b) a large number of small pores
   - (c) a sufficient number of large pores
   - (d) plenty of stones?

Exercise 10.3

Sand grade investigation

Background

Often it is useful to use sand as a surface material either on which to stand container-grown plants, or as a root zone mixture for sports turf and lawn topsoils. In all these situations a special blend of different sand grades should be used. The resulting 'mixture' should have enough large pore spaces to allow water to drain away, but enough small pores to transport water by capillary action (siphoning) up to the plant roots. Once such material is called 'Efford' sand, designed by Efford environmental research station as a standing material for container-grown plants. Other materials that you may like to investigate include the Texas University specification for the United States Golf Association (USGA), golf green root zone mixture.

Aim

To observe the substantial variations in sand grades present in Efford sand.

Apparatus

Efford sand

sieve nest

measuring cylinders (10, 100 and 500 ml)
500 ml beaker

Method

1. Measure out approximately 500 ml of sand.
2. Arrange sieve nest in the following order: receiver tray, 0.06 mm (fine), 0.2 mm (medium), 0.6 mm (coarse), 2 mm (stones), lid.
3. Place 500 ml of sand onto the top sieve (2 mm) and shake for 5 minutes.
4. Separate sieve nest.
5. Place each grade of sand into a measuring cylinder and record its volume.
6. Calculate the percentage of each grade within your 500 ml sample volume.
7. Save some of each grade sand in a weigh boat for use in Exercise 10.2.

Results

Enter your results in the table provided.

Particle sizes	Volume within 500 ml sample (A)	% of sand grade within 500 ml sample = A/500 ml ×100
Coarse sand 0.6–2 mm		
Medium sand 0.2–0.6 mm		
Fine sand 0.06–0.2 mm		
Receiver tray <0.06 mm		

Conclusions

1. What grade of soil falls through all the sieves onto the receiver tray?
2. What properties does Efford sand possess which are useful in horticulture?
3. Which of these statements explains how in spring a heavy soil warms up more slowly than a sandy soil:
   (a) it becomes colder during the winter period
   (b) it has a lower water-holding capacity
   (c) surplus moisture does not drain away so quickly
   (d) warm air cannot enter so easily?

Exercise 10.4

Soil texture assessment

Background

The mineral fraction of the soil is the collective term for sand, silt and clay particles. The relative amounts of these particles is defined as soil texture. The ability to accurately determine the soil texture is a skill which can be particularly challenging but which will pay great dividends when properly mastered. Many apects of land management rest upon its determination in the field situation. For example, pesticide and herbicide application rates, nitrogen fertilizer recommendations and general skills of manipulating the growth medium to enhance plant growth and development rest upon the foundation of soil texture.

The ability to determine soil texture by hand assessment is a skill which will require regular practice. It is a good idea to build up a collection of soils with different textures and repeatedly practise on these samples. The purchase of a local soils map from the National Soils Resources Institute will be a good starting point. A more accurate determination of soil texture may be obtained by submitting a soil sample to a laboratory for analysis. However, the hand texture assessment method will be adequate in most cases. It is suited for most soils, except those with above 45 per cent organic matter, which are termed 'organic' soils rather than mineral soils.

Aim

To develop skills to correctly conduct a hand texture assessment on different soil types.

Apparatus

Figure 10.3
Figure 10.7
soils of different textures labelled 1–10
water bottle

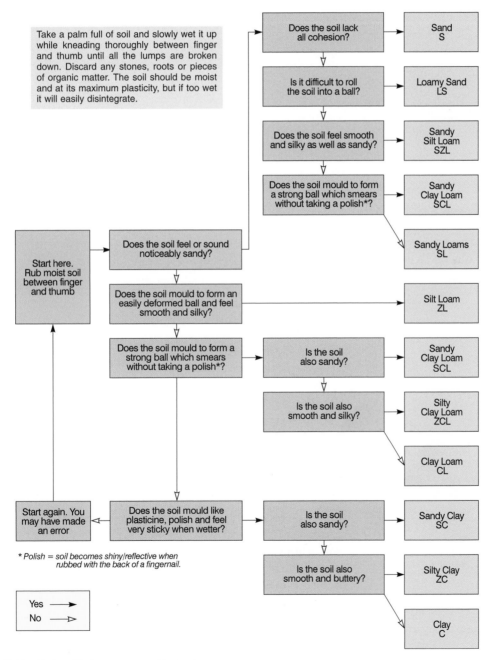

Figure 10.7 Field guide to soil texture assessment by hand

Useful website

www.silsoe.cranfield.ac.uk/nsri

Method

Using Figure 10.3 assess the textures of the soils present in the sample buckets.

Results

Enter your results in the table provided.

Soil sample number	Texture assessment determination
1	
2	
3	
4	
5	
6	
7	
8	
9	
10	

Conclusions

1. How many textural classes are there?
2. Define the term 'soil texture'
3. Explain the relationship between soil texture and temperature in spring.

Exercise 10.5

Soil texture assessment of personal plot sites

Background

Now that some practice has been gained at the principles of determining soil texture it will be necessary to apply these techniques to a field situation. This may be to investigate the soil from a garden, demonstration plot, allotment, sports field, cricket wicket, golf green or other area in which you have a personal interest.

Aim

To determine the texture of site top and subsoils using the hand texture assessment method and link your findings to the subsequent management practices needed.

Apparatus

Figure 10.7
Plot's topsoil
Plot's subsoil

Method

Using Figure 10.7 assess the textures of the soils in your chosen site.

Results

Enter your results in the table provided.

Plot sample	Texture assessment
Topsoil	
Subsoil	

Conclusions

Comment on how the soil texture of your site will effect the management of it in the following areas:

(a) water-holding capacity
(b) aeration
(c) temperature at different times of year
(d) nutrient availability and need for fertilizer
(e) need for organic matter or mulches
(f) drainage
(g) workability.

Answers

Exercise 10.1. Properties of soil particles

Results

Sand feels gritty, similar to table salt

Silt feels silky/soapy, similar to butter

Clay feels sticky, similar to Plasticine or putty

Conclusions

1. (a) 0.6–2.0 mm (600–2000 μm)
    (b) 0.2–0.6 mm (200–600 μm)
    (c) 0.06–0.2 mm (60–200 μm)
    (d) 0.002–0.06 mm
    (e) 0–0.002 mm

2. (a) Sand: prone to drought, well drained, well aerated, low water-holding capacity, low nutrient status, referred to as hungry, early warm and light soils.
    (b) Silt: deep soils, cake drains, low nutrient and water content, slow drainage, low aeration.
    (c) Clay: high nutrient content, high water content, warm up slowly in spring, poorly drained, referred to as cold, late and heavy soils.

3. (a) sticky.

4. Clay has a high nutrient content and high available water content.

5. Sand feels gritty. Advantages are well drained and well aerated. Disadvantages are prone to drought, low water-holding capacity and low nutrient status.

Exercise 10.2. Features of course, medium and fine grade sand

Results

Coarse sand feels like caster sugar; easily audible

Medium sand feels like table salt, audible

Fine sand feels like ground pepper; slightly audible.

Conclusions

1. Medium sand; it provides a range of small and large pore spaces.

2. Coarse sand; the texture cannot easily be changed; however coarse sand will provide extra aeration and large pores.

3. The organic matter will quickly be composed by soil organisms including fungi and bacteria. For this reason they are referred to as 'hungry' soils.

4. (c) A sufficient number of large pores. Large pores provide drainage and aeration. Small pores can remain filled with water and 'irrigate' plants through capillary action.

Exercise 10.3. Sand grade investigation

Results

Efford sand is composed of approximately:

  coarse sand: 30–45%
  medium sand: 40–60%
  fine sand: 5–15%

Conclusions

1. Particles less than 0.06 mm, typically either crushed minerals or particles of silt and clay.

2. It possess sufficient large pores to ensure good drainage and a lack of waterlogging but sufficient medium and small pores to ensure adequate plant watering through capillary action. Plants grown of Efford sand beds are of a significantly higher quality than those raised on traditional standing materials such as mypex with overhead irrigation.

3. (c) Surplus water cannot drain away so quickly and since water requires energy to heat up the soil remains significantly colder than sandy soils.

Exercise 10.4. Soil texture assessment

Results

A broad range of known soils should be investigated using samples collected using a soils map as a guide to texture.

Conclusions

1. There are 11 textural groups noting that SCL appears twice in Figure 10.7.

2. Soil texture is the relative percentage of particles of sand, silt and clay present in the soil.

3. Sandy soils are light and airy, warm air enters through the pores easily in spring and early germination results. Heavier soils containing more clay have smaller pores and retain water which requires energy to warm up; consequently they reach a suitable temperature for seeds to germinate much later in the season than sandy soils.

Exercise 10.5. Soil texture assessment of personal plot sites

Results

Results will be dependent on specific soils sampled.

Conclusions

Appropriate sensible comments should be made in each of the catergories (a) to (g), based upon the texture identified.

Chapter 11   Soil structure and profiles

Key facts

1. Soil structure refers to the binding of soil particles together into peds and clods.
2. A soil profile is a description of the characteristics of soil horizons, often separated into topsoil and subsoil.
3. Common soil types in Britain are rendzina, brown earths and podsols.
4. There are seven soil structure types; single grain, crumb, blocky, prismatic, platy, columnar and massive.
5. A soil profile pit is usually dug to a depth of 1 m and all the horizons described.

The study of soil in the field

A horticulturist may think of soil as only the top 30 cm of material under cultivation. A civil engineer will think of soil as several metres of unconsolidated material overlying solid bedrock. However, a soil scientist will classify soil as the material formed by the action of some organic agency on the original rock, or parent material. In Britain this usually involves the study of the top 1–2 m of soil overlying unaltered parent material. In the tropics by comparison, ten times this depth may be required to include the whole soil.

The cross-section of the soil seen on the face of a pit, quarry or other cutting is called the soil profile. It consists of layers of material (soil horizons), each of which appears to have different properties from the horizons above and below. These horizons are the product of soil formation processes or 'pedogenesis', and should be distinguished from the layering simply inherited from the stratification of the parent material.

A soil profile is a vertical slice through the soil, usually obtained by digging an examination pit 1 m deep. It may be simply divided into topsoil and subsoil as in Figure 11.1.

Litter layer	Characteristics of a good management	
Top soil 0–15cm	• crumb structure • highly fertile • lots of soil organisms	• high organic matter content • lots of root development • freely draining
Subsoil 15–30cm	• coarse structure • low organism populations • low organic matter content • freely draining	
Parent material 30+ cm	• underlying rock geology structure	

Figure 11.1 Simple soil profile

The study of the properties of these horizons may tell us a considerable amount about the characteristics of the particular soil and its agricultural or horticultural potential. For a detailed assessment of properties, including structure, a small hole should be dug to reveal the soil profile. The soil is then examined to a depth of 1 m.

Several soil profile types occur in Britain. Three of the commonest are rendzina, brown earth and podsol. Figure 11.2 shows characteristics of these three types. A rendzina is a very shallow soil overlying chalk and is common on the chalk downlands of southern England. Because they overly chalk, these soils are very limey. A brown earth is a deep

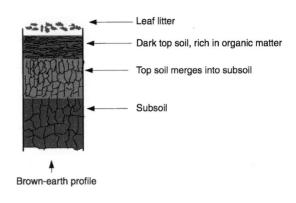

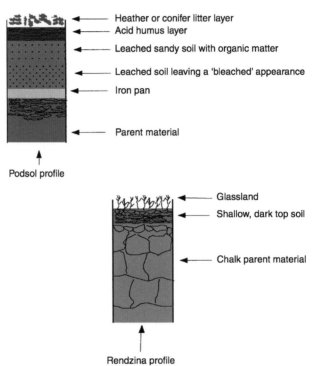

Figure 11.2 Common soil profile

fertile soil common across Britain. Like podsols, they often suffer from leaching and have neutral to acid pH. Generally these soils are well-drained. The brown colour comes from mixing the iron and aluminium oxides throughout the soil through moderate leaching. The name podsol comes from the Russian word meaning 'ash-like'. A podsol is a soil with sandy surface horizons which have been severely leached, washing down free iron and aluminium oxides. This results in the formation of an iron-pan below and a bleached sandy horizon above. They have very easy to see, distinct soil horizons. Podsols are common in the northern hemisphere, especially deciduous forest soils.

The exercises in this chapter are designed to introduce you to soil profile structures and to enable you to interpret your observations to make recommendations on how the soil should be improved/managed to achieve good plant growth. The following areas will be investigated:

- soil structure
- general background to site under investigation

- procedure for digging a soil inspection pit
- examination of soil profile pit
- production of a record sheet
- conclusions and horticultural significance of the findings.

Soil structure

The soil structure refers to the binding together of soil particles into natural (peds) or artificial (clods) structural units of different shapes and sizes.

There are seven different structural units that can be easily recognized in the field. These are illustrated in Figure 11.3.

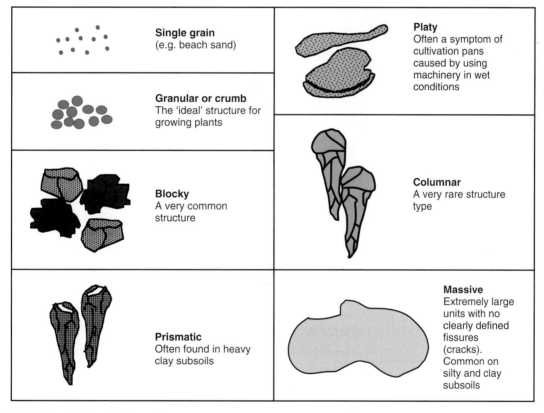

Single grain
(e.g. beach sand)

Granular or crumb
The 'ideal' structure for growing plants

Blocky
A very common structure

Prismatic
Often found in heavy clay subsoils

Platy
Often a symptom of cultivation pans caused by using machinery in wet conditions

Columnar
A very rare structure type

Massive
Extremely large units with no clearly defined fissures (cracks). Common on silty and clay subsoils

Figure 11.3 Soil structure types (not to scale)

Figure 11.4 An ideal 'crumb' soil structure

Structure units may be further divided into very coarse, coarse, medium and fine size units. An easy way to become familiar with structure units is to go out into the garden and collect a small handful of soil from the soil surface. Compare the shapes with those in Figure 11.3. Another technique is to dig a spade full of soil and turn it over in the air before allowing the soil to fall off the spade and hit the ground. You should find that the soil will fracture along its natural fissure lines as it breaks up into its structural units on impact.

Exercise 11.1

Soil structure and profile assessment

Aim

To act as an introduction to soil profile structures; to interpret observations from the assessment; and to make recommendations on how the soil should be improved/managed to achieve good plant growth.

Apparatus

Metre rule	Clipboard	Paper
Pencils	Spade	Plastic bags
Sample bags	Labels	Forks
Augers	Trowel	Wash bottles
Buckets	BDH pH kits	

Figure 11.5 Soil profile pit

Method

Select a site for investigation. Figure 11.5 shows an example soil profile pit. A description of a soil profile involves the following steps.

Stage I: General background

Collect the following information:

1. A description of the site with map reference if available.
2. Details of vegetation.
3. Topography and drainage (e.g. shape of the land, undulating, rolling, flat, sloped etc.).
4. Local geology.
5. Signs of erosion.
6. Climate data (e.g. rainfall).

Stage II: Production of profile inspection pit

The procedure when digging and filling a soil profile pit is as follows:

7. Choose a position away from gateways and tracks
8. Dig a pit approximately 1 m × 1 m × 1 m deep.
9. When digging the inspection pit care should be taken to:
   (a) cut the turf so that it can be replaced easily and neatly
   (b) keep the topsoil and subsoil separate so that each can be returned to its own position without mixing them
   (c) **clearly mark** were holes are – use a long stick or cane with a cloth, flag or rope the area off.
   (d) regularly consolidate the soil when refilling the pit
   (e) replace all the soil in the hole; this will leave the site 'proud' and allow for settlement.
10. While digging:
    (a) note hardness and compactness of the different horizons
    (b) check for stones or pans if spade hits resistance.
11. Do not leave unmarked pits in fields. If, for any reason, you wish to leave a pit unattended you must inform the land owner and obtain their permission.

12. **Never** leave pits unattended in fields containing livestock or in gardens where children have access.

13. When opening a profile pit in an orchard, remove the top 2–3 cm of soil (the herbicide layer) and keep it separate. It should be returned to the top of the soil after the profile pit has been filled in.

Stage III: Examination of soil profile pit

14. Clean a face of the pit for inspection.

15. Note the number of different horizons in the soil, the size of each horizon and the depth at which they occur.

16. Pattern of root development and penetration – note signs of horizontal rooting, deformed roots, frequency of fine roots, presence of roots within or between structure units, etc.

17. Note depth and frequency of worm channels.

18. Note points where water seeps into the pit.

19. Take samples of topsoil and subsoil for chemical analysis.

20. For each horizon estimate the following characteristics, and record on your result sheet:

Horizon designation (topsoil/subsoil)

Horizon depth and thickness

Colour

Texture (use Figure 10.7)

Structure (blocky, crumb, platy, prismatic, single grain, massive)

Pans

Concretions (manganese, calcium carbonate, etc.)

Mottling (orange rust colours, indicating seasonal waterlogging), see Figure 11.6

Gleying (bluey-grey colour, indicating permanent waterlogging)

Fissures: these are natural cracks in the profile.

Figure 11.6 Signs of mottling and gleying

Fissures	Porosity (implied from fissure size)
Very fine, < 1 mm wide	(very slightly porous)
Fine, 1–3 mm	(slightly porous)
Medium, 3–5 mm	(moderately porous)
Coarse, 5–10 mm	(very porous)
Very coarse, > 10 mm	(extremely porous)

Stones	Abundance	Size
	Stoneless (<1%)	Very small stone (2–6 mm)
	Very slightly stony (1–5%)	Small stones (6 mm–2 cm)
	Slightly stony (6–15%)	Medium stones (2–6 cm)
	Moderately stony (16–35%)	Large stones (6–20 cm)
	Very stony (36–70%)	Very large (20–60 cm)
	Extremely stony (> 70%)	Boulders (>60 cm)

Organic matter (roots and animals)

pH: use a BDH pH testing kit or similar item (see 'Measuring soil pH' exercises)

Any other relevant information (e.g. a reclaimed rubbish dump, previously used for cattle, small garden subject to excessive dog urea, etc.).

Results

Enter your results in the blank 'Soil Profile Examination Record Sheet' (Figure 11.7). A completed example result sheet is included for reference (see Figure 11.8).

Soil Structure and Profile Results Sheet

Name: _____ Date: ___ / ___ / ___

Location: _____ Crop: _____

_____ _____

Depth cm	Sketch	Profile description
0		
5		
10		
15		
20		
25		
30		
35		
40		
45		
50		
55		
60		
65		
70		
75		
80		
85		
90		
95		
100		

Figure 11.7 Soil profile examination record sheet

Name: Pedologic Landscaping   **Date:** 04.11.02

Location: Plot 1, Sandy Lane project   **Crop:** Mixed trees and shrubs
Gardner's City
Bloomingtown

Depth Sketch   **Profile description**
cm   Surface – Light brown silt loam, Angular blocky. Very small stones 6–15%, slight capping pH 6.0–6.5

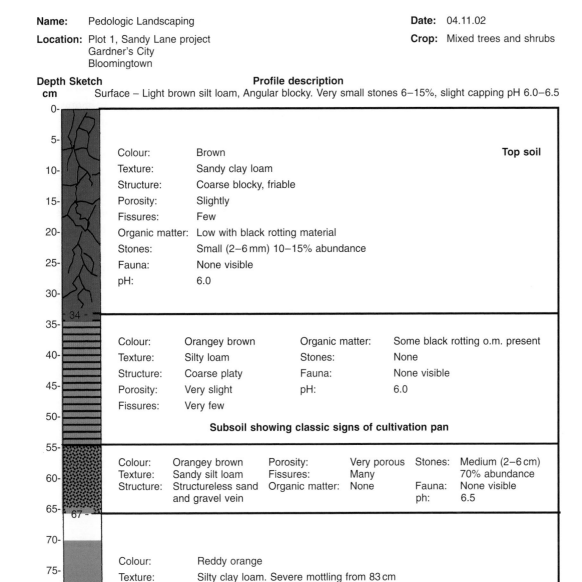

			Top soil
Colour:	Brown		
Texture:	Sandy clay loam		
Structure:	Coarse blocky, friable		
Porosity:	Slightly		
Fissures:	Few		
Organic matter:	Low with black rotting material		
Stones:	Small (2–6 mm) 10–15% abundance		
Fauna:	None visible		
pH:	6.0		

Colour:	Orangey brown	Organic matter:	Some black rotting o.m. present
Texture:	Silty loam	Stones:	None
Structure:	Coarse platy	Fauna:	None visible
Porosity:	Very slight	pH:	6.0
Fissures:	Very few		

Subsoil showing classic signs of cultivation pan

Colour:	Orangey brown	Porosity:	Very porous	Stones:	Medium (2–6 cm)
Texture:	Sandy silt loam	Fissures:	Many		70% abundance
Structure:	Structureless sand and gravel vein	Organic matter:	None	Fauna:	None visible
				ph:	6.5

Colour:	Reddy orange
Texture:	Silty clay loam. Severe mottling from 83 cm
Structure:	Structureless massive. Very compacted
Porosity:	None
Fissures:	None
Organic matter:	Some black rotting material
Stones:	None
Fauna:	None visible
pH:	6.5

Figure 11.8 Example of soil profile examination record sheet

Conclusions

A report containing observations, deductions and management decisions should be prepared. A diagram of the profile should be draw and interpreted for management in terms of:

- ease of cultivation
- soil water-holding capacity

- drainage
- aeration works and subsoiling
- stability of topsoil.

The report should also contain suggestions for possible planting for the site examined along with reasons for this choice. Suggestions to improve the site for planting may also be included. Before finally deciding on the plant to be grown, the nutrient status of the soil should be considered and a fertilizer recommendation made.

The report should also contain suggestions for possible planting for the site examined along with reasons for this choice. Suggestions to improve the site for planting may also be included. Before finally deciding on the plants to be grown, the nutrient status of the soil should be considered and a fertilizer recommendation made.

Interpret the significance of your findings to horticulture by completing comments below.

1. What are the problems with this soil?
2. What are the benefits from this soil?
3. What is the management practice required to improve this soil?
   (a) cultivation
   (b) soil water-holding capacity
   (c) drainage
   (d) aeration works and subsoiling
   (e) stability of topsoil
   (f) other.
4. Name three suitable plants/crops for use uses on this site.
5. Why are the topsoils and subsoils kept separately?
6. Explain what mottling and gleying tell us about the soil.
7. Define the term soil 'structure'.
8. Distinguish between soil texture and soil structure.

Answers

Exercise 11.1

Results

A profile diagram 5 early identifying and describing horizons similar to Figure 11.8 should be produced.

Conclusions

1. A commentary on the profile should be given including plants that are suitable/unsuitable together with comments on: ease of cultivation, water-holding capacity, drainage, aeration works and topsoil stability.

2. As above.

3. As above.

4. As above.

5. Topsoil is rich in organic matter and plat nutrients. It is kept separate from the relatively poor quality subsoil so that they can be reinstated in the same position.

6. Mottling is indicative of semi-permanent or seasonal waterlogging. This shows itself as orangey-red colourings in the soil. Gleying is indicative of permanent wetness and shows itself as bluey-grey discolouration.

7. Soil structure refers to the binding together of soil particles into natural (peds) or artificial (clods) structural units of different shapes and sizes such as granular and blocky.

8. Soil texture refers to the relative percentage of particles of sand, silt and clay present in the soil such as sandy clay loam or silty clay. Soil structure refers to the binding together of soil particles into natural (peds) or artificial (clods) structural units of different shapes and sizes such as granular and blocky.

Chapter 12   Soil water

Key facts

1. Soil water exists in different forms including available, unavailable and drainage water.
2. Soil water is held too tightly by particles to be liberated by plants, and plants will wilt if they do not receive additional watering.
3. Soil water is held around the surface of soil particles. Thus the soil texture with the largest surface area has the greatest available water content.
4. Waterlogged conditions lead to anaerobic respiration and the production of toxic alcohol as a waste product.
5. Waterlogged soil often has a rotten egg smell generated from the production of hydrogen sulphide gas.
6. Water can move through soil by capillary action.
7. Many diseases are associated with waterlogged soils including Botrytis and damping-off.

Background

The soil is composed of solid particles, air spaces and water, all of which are essential for healthy plant growth. However, different soil texture groups will have varying proportions of these properties which may influence the growth rate of plants. Variations in particles sizes, for instance, greatly modify the number and size of pore spaces in the soil and will subsequently influence the ability of the soil to drain freely. With predicted change in global climates conservation of soil water is considered of ever-increasing importance. A primary purpose of soil is to store water for use by plants. Soils will lose moisture through evaporation from the soil surface and evapo-transpiration from the plant.

If soil moisture reserves become low, plants will become stressed and begin to wilt (temporary wilting point). If the interval between rainfall increases, permanent wilting point is reached and the plant will die. With climate change predicted more consumers are turning to drought-tolerant plants combined with soil conservation measures.

The exercises in this chapter will investigate aspects of the water content of a variety of soils and also the consequential response of plants to over- and under-watering.

Soil water can exist in different forms in a soil profile (see Figure 12.1). Some simple definitions may be useful.

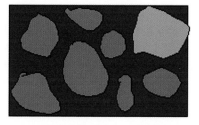

Saturation point

All pore space filled with water. No air present

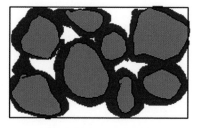

Field capacity

Large pores have drained of water. Small pores hold maximum amount of water against the pull of gravity

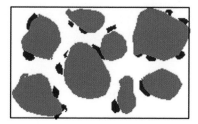

Permanent wilting point

Large and small pore filled with air. Water held too tightly to soil particles to be liberated by plant roots. Plants will wilt and die

Figure 12.1 Soil moisture conditions

Saturation point: All soil pores are full with water (waterlogged).

Field capacity: Large pores are drained of water but small pores around particles contains films of water. This is the maximum amount of water that a soil can hold after the excess has drained away under the force of gravity.

Drainage water: The volume of water between saturation point and field capacity, which drains under the force of gravity (also called gravitational water).

Available water: Water in the soil which plants can take up, held between field capacity and permanent wilting point.

Permanent wilting point: The water content of a dry soil at which the water is held by the soil particles too tightly to be released by plant roots. Plants are therefore unable to draw water. Irreversible wilting follows and the plant will die.

Exercise 12.1

Available water-holding capacity

Background

The available water-holding capacity (AWHC) is a measure of soil available water content in grams per cubic centimetre ($g\,cm^{-3}$), and is often expressed as a percentage. This water is available for plants to draw up. It is the difference between the soil water content at field capacity and wilting point water. Water is held on the surface of soil particles. Therefore the soil with the largest surface area will have the greatest available water-holding capacity. In practice this means that clay soils hold more water than sandy soils and will require less irrigation or watering in the hot summer weather. For this reason clay is often used as an ingredient in container compost, particularly where the interval between watering is likely to be great. However, they may require help to drain excessive winter rainfall away.

Aim

To determine the available water-holding capacity of a soil at field capacity of sand and clay soils (approximate method).

Apparatus

Sandy soil	Clay soil
Polystyrene beakers (2)	Marking pen
Measuring cylinder	Filter paper
200 ml beakers (2)	Reservoir (washing up bowl)
Pen knife	Analytical balance

Useful websites

www.swcs.org

www.physicalgeography.net

Method

You may find it helpful to produce a labelled diagram of the apparatus. Record all readings in the relevant place in the results table (letters in brackets refer to rows in table).

1. Using the balance determine the weight (a) of an empty polystyrene beaker.
2. Mark two polystyrene beakers with your initials and soil type (sand or clay).
3. Calculate the volume (b) of the polystyrene beaker using the following procedure:
   (i)   Fill beaker with water.
   (ii) Pour contents into measuring cylinder.
   (iii) Record volume of beaker (b) in ml.
4. Fill one cup with dry sandy soil, firming regularly.
5. Fill second cup with dry clay soil, firming regularly.
6. Weigh each filled cup on the analytical balance. Record the result in row (c).
7. Calculate the weight of soil in the breaker, recording the results in row (d) ((c)–(a)).
8. Pierce bottom of each beaker with a knife to create drainage holes.
9. Bring each soil to saturation point by slowly lowering beakers into water reservoir, allowing the soil to become wetted from below. All the pores should now be filled with water.
10. When saturated,(i.e. water floats on surface of soil and all pores spaces are full), gently place a piece of folded filter paper against the bottom of each beaker to prevent water draining away, and quickly transfer beaker into a 200 ml collecting beaker, and allow to drain freely.
11. When drainage stops record the volume of drainage (gravitational) water held in the collecting beaker (e). Your soil is now at field capacity.
12. Weigh beaker and soil at field capacity (f).
13. Determine the weight of soil at field capacity (g) ((f)–(a)).
14. Determine the weight of available water content held at field capacity (h) ((g)–(d)).
15. Calculate the available water-holding capacity by dividing the weight of available water content (h) by the sample volume (b), noting that the density of water is $1.0\,g\,cm^{-3}$ (cm^3 = 1 ml., i.e. 1 g of water $\approx$ 1 ml volume).
16. Express your finding as a percentage available water-holding content (j) using the formulae:

$$\% \text{ available water-holding capacity} = \frac{\text{available water content weight (h)}}{\text{volume of soil contained in beaker (b)}} \times 100$$

Results

Enter your results in the table provided.

Conclusions

1. Which soil type had the largest available water-holding capacity?
2. Explain your answer in Question 1.
3. How might the results have varied for a silt loam textured soil?
4. Which soil became wetted up quicker? And why?
5. Which soil will require more frequent watering?
6. Explain which soil is more likely to be draught-resistant in the summer months.
7. How will the variations in available water-holding capacity influence the establishment or regrowth of plants in the spring months?
8. Describe how the suitability of these soils would influence the choice of plants to be grown.

Results

Enter your results in the table provided.

Measurements	Example soil	Soil A	Soil B
(a) Weight of empty polystyrene beaker (g)	20 g		
(b) Volume of polystyrene beaker (ml)	250 ml		
(c) Weight of beaker + soil (g)	265 g		
(d) Weight of soil (c) − (a) (g)	265 − 20 = 245 g		
(e) Volume of drainage water (ml)	50 ml		
(f) Weight of soil + beaker at field capacity (g)	315 g		
(g) Weight of soil at field capacity (f) − (a) (g)	315 − 20 = 295 g		
(h) Available water content (g) − (d) (g)	295 − 245 = 50 g		
(i) Available water-holding capacity (h/b) (g cm^{-3})	50/250 = 0.2 g.cm^{-3}		
(j) % available water-holding capacity (h) ×100/(b)	50 × 100/250 = 20%		

Exercise 12.2

Effects of waterlogged soils

Background

There are many reasons why a plant may come to be growing in waterlogged soils. It may be due to compaction panning preventing free drainage or as a result of overwatering. The effects on plant growth (physiology), can be dramatic. Plants may appear stunted, roots may die off and in the absence of oxygen, anaerobic (without oxygen) respiration may occur, producing toxic alcohol as a waste product (see Chapter 6). This may be accompanied by the production of sulphur dioxide gas, producing a rotten egg smell. There are also many diseases carried by water including botrytis and damping-off.

Aim

To demonstrate the negative effects on plant growth of waterlogged soils.

Apparatus

Seed trays (2)
Oil seed rape seed
Clay loam soil
Water

Useful website

www.appliedresearchforum.org.uk

Method

You may find it helpful to draw a labelled diagram of the plant appearances.

1. Prepare both seed trays with soil and sow with suitable seeds.

2. Water Tray A normally.
3. Over-water Tray B.
4. After 7 days record your results.

Results

Enter your results in the table provided.

Treatment	No. of seeds sown (X)	No. of successful germinating seeds (Z)	% germination = Z/X × 100	Physiological appearance of roots, stems, leaves and flowers
Tray A Normal watering				
Tray B Overwatering				

1. Calculate the percentage germination from each treatment.
2. Describe any physiological differences between the two groups of seedlings.

Conclusions

1. Explain the differences in your observations.
2. Identify any diseases present on the waterlogged specimens.
3. Name another disease common in waterlogged conditions.

Exercise 12.3

Capillary rise

Background

Water moves through the spaces (pores) between soil particles by cohesion, surface tension and adhesion. This is called capillary action. As a plant absorbs water through the roots it creates a space into which other water will flow rather like a siphon. Water may also evaporate from the surface of the soil, drawing more water upwards from the subsoil also by capillary action. The size of the soil pores will influence the speed with which capillary action works and also the ability of the soil to drain effectively.

Aim

To demonstrate the process of capillary rise (water movement through soil).

Apparatus

Capillary tubes (3)
Water
Retort stand
Blended sand grades (course, medium and fine) (3)
Tray

Method

You may find it helpful to draw a diagram of the apparatus.

1. Fill each tube with a different grade of sand (see Figure 12.2).
2. Clamp each tube in a retort stand.

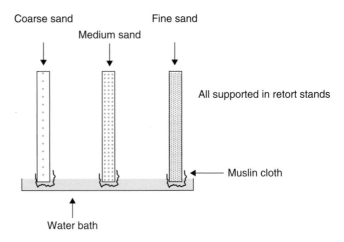

Figure 12.2 Effect of sand grades on capillary action

3. Place tubes in a tray of water.
4. Observe the reaction.

Results

1. Measure the water level reached in each tube and record your result in the table provided.
2. Explain the variations in the readings obtained.

Capillary tube	Height of wetting front (mm)	Explanation of variations
Tube A		
Tube B		
Tube C		

Conclusions

1. What process is responsible for the rise in the level of the water in the three tubes?
2. Define the term 'capillary rise'.
3. What significance of your finding for drainage of horticultural land?
4. Describe the significance are your finding for the irrigation of plants.
5. Describe the influence of capillary rise on watering of container-grown plants.

Exercise 12.4

Plant physiological response to water stress

Background

Plants will vary in their response to different degrees of water stress.

Waterlogged soils tend to produce shrivelled leaves and roots. The rotten egg smell associated with sulphur dioxide gas produced in anaerobic conditions is also a common symptom. Drought conditions often result in a flush of root growth as plants actively seek out new sources of water, followed by leaf fall and plant death. Usually, plants that are semi-submerged in water respond by their roots growing upwards away from the waterlogged conditions. Container-grown house plants that have been over watered often demonstrate these characteristics. This is because aerobic conditions are essential for healthy growth and development. These exercises will investigate some of these characteristics.

Aim

To observe the effects of: (a) waterlogging the root zone, (b) half waterlogging the root zone, and (c) normal watering.

Apparatus

> *Pelargonium* pot plants (3)
> Reservoirs (e.g. bucket) (3)
> Water

Method

1. Stand Plant A in a reservoir with the water completely submersing the route zone (waterlogged).
2. Stand Plant B in a reservoir with water filled to half the depth of the pot.
3. Stand Plant C in a reservoir and water normally.
4. Observe each specimen over the next 7 days.
5. Examine the leaves, stems, flowers and root balls.
6. State how their outwards appearance varies.

Results

Enter results in the table provided.

Treatment	Roots	Stem	Leaves	Flowers
Plant A (waterlogged)				
Plant B (semi-waterlogged)				
Plant C (normal irrigation)				

Conclusions

1. Describe the physiological reaction of plants to waterlogging.
2. Explain the conditions that favour the production of hydrogen sulphide gas.
3. Give two management practices in horticulture that might lead to semi-waterlogged conditions.

Answers

Exercise 12.1. Available water-holding capacity

Results

Sandy soil is about 20 per cent AWHC under this semi-quantitative method. A clay soil typically comes about at 55 per cent AWHC.

Conclusions

1.   Clay had the largest AWHC.

2.   Clay particles are very small (less than 0.002 mm), consequently they have a larger surface area per unit volume than sandy soils. Since water is held on soil particles the clay holds more than the sand. The concept of surface area has proved difficult for students to grasp. Sometimes using models helps to convey the information. For example consider a dustbin filled with footballs and another filled with ping-pong balls. The bin with ping-pong balls represents the smaller clay particles which, after adding up the total surface area of all the ping-pong balls, has a greater surface area than the bin with footballs.

3.   Silt loam contains less than 20 per cent clay, less than 20 per cent sand and between 85 and 100 per cent silt. Silt is an intermediate particle size (0.002–0.06 mm) and will give a result between sand and clay of about 35 per cent AWHC.

4.   Sand becomes wetted-up quicker because water moves through the large pores easily and because water does not adhere to the particles as easily as clays. The clay soil has small pores and very absorbent particles. A characteristic of clay is that it swells when wet. Thus the water moves slowly through the pores, wetting up particles slowly.

5.   Sand will require more frequent watering due to the large pore sizes, creating a free draining soil and lower AWHC.

6.   Clay will be more drought-resistant because it has a higher AWHC.

7.   Clay soils are cold and late. They hold on to water well and in the spring this tends to result in lower soil temperatures. Sands have large pores, creating air spaces into which the warmer spring air can flow, warming it up more speedily and stimulating growth. For this reason sandy soils are often called *warm* or *early* soils.

8.   Sandy soils require regular irrigation. Plants that are adapted to survive in drought conditions would be suitable, e.g. rosemary. Alternatively grow plants which have deep roots able to exploit a large soil volume from which to draw water. Clay soils require soil additives such as farmyard manure, to break up the structure. They are susceptible to waterlogging. Either plants need to be grown on raised beds or plants able to withstand water stress should be planted. Most conifer trees would be suitable.

Exercise 12.2. Effects of waterlogged soils

Results

Tray A: up to 100 per cent germination. Normal physiological appearance.

Tray B: 0–60 per cent germination. Those that do germinate suffer from damping-off and have stems 'pinched-in'.

Conclusions

1.   Waterlogged soil create both anerobic conditions which are toxic and stimulates waterborne diseases such as damping-off which kills the seedlings.

2.   The disease would be damping-off, evidenced by a pinched-in stem.

3.   *Rhizoctonia* root rots and *Phytophthora*.

Exercise 12.3. Capillary rise

Results

The order in which water is raised to the greatest height is fine, medium and course sand respectively, due to the variation in pore sizes affecting capillary action.

Conclusions

1.   Capillary action

2.   Water moving through the pores between soil particles by surface tension from a wetter area to a drier area.

3. The larger the soil pores the better the drainage.

4. Capillary action is stronger for fine sand than course sand. Sand beds (e.g. Efford bed) are often used to grow plants.

5. A good sand bed will supply a reservoir of water for plant growth but also enable free drainage of surplus rainfall. Capillary action is achieved by constructing the sand bed of a range of particle sizes to enable water to move through a vertical height of 10–23 cm, although the depth of the bed may only be 3–5 cm. Efford sand beds are composed of approximately 30–45 per cent course sand, 40–60 per cent medium sand and 5–15 per cent fine sand.

Exercise 12.4. Plant's physiological response to water stress

Results

Plants will show increasingly pronounced physiological responses to the water stress. This will include shrunken leaves, dead roots, hydrogen sulphide smells and premature flower death.

Conclusion

1. Describe as given in results.

2. Waterlogged soils creating anaerobic conditions.

3. Compacted soils and poor film in hydroponic production.

Chapter 13   Measuring soil pH

Key facts

1. pH is the negative logarithm of the hydrogen ion (H⁺) activity. This is a measure of how acid or alkaline the soil is on a scale of 0 to 14.
2. The ideal pH for most plants in soil is pH 6.5 and pH 5.8 in compost.
3. Calcifuge species prefer acid soils, e.g. heather.
4. Calcicole species are lime-tolerant, e.g. some alpines and Prunus.
5. There are several ways to measure pH. Labs will use a complex pH electrode but the test tube method, using universal indicator solution, is a widely used field technique.

Background

The importance of pH

pH is used to measure the acidity or alkalinity the growth medium. Plants grown in soils with inappropriate pH will fail. For example, the public are fond of purchasing heathers and other acid-loving species, planting them in alkaline soils and complaining when they grow poorly or die. In addition, the availability of nutrients changes with pH, becoming either deficient or toxic in the extremes. The pH at which most nutrients are freely available for absorption by plant roots is pH 6.5 in soils and pH 5.8 in composts. Similarly, there are many plants grown in commercial crop production that will only tolerate a narrow band of acidity. Accurate pH management is therefore vital to ensure success.

As horticulturalists we need to know:

- the importance of pH
- methods of measuring soil pH
- specific plant pH preferences
- methods to correct soil acidity by liming (raise pH)
- methods to increase soil acidity by sulphur (lower pH).

The pH scale

pH is measured on a **log scale** from 0 to 14, as illustrated in Figure 13.1. pH 7.0 is neutral (neither acid or alkaline); below, pH 7.0 is acidic and above, is alkaline.

The pH scale. pH is measured on a **log scale** from 0–14:

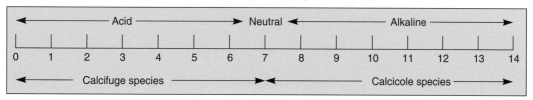

Figure 13.1 The pH scale

Plants vary greatly in their tolerance to acidity levels. Lime-loving (alkaline-loving) plants are called **calcicole** species. Lime hating (acid-loving) plants are called **calcifuge** species. A list of plant pH preferences is given in Table 13.2.

pH and Hydrogen ion (H$^+$) concentration

Technically, pH is defined as the negative logarithm of the hydrogen ion concentration.

pH counts protons of hydrogen ions (H$^+$). It only measures those hydrogen ions free in the soil solution. Protons that are bound to molecules are not counted. However, there is a continual slow release of hydrogen ions into the soil solution from other molecules. In addition, some hydrogen ions leave the soil solution to bind with chemical molecules. Thus there is a dynamic equilibrium where each hydrogen ion is free in solution for a very short time of approximately 2×10^{-12}

seconds. pH measures this trading activity in and out of the soil solution. A neutral solution has a hydrogen ion concentration of 10^{-7} (moles per litre). The significance of this log scale is that, for example, pH 5 is ten times more acidic than pH 6, but one hundred times more acidic than pH 7, and will therefore require substantially more lime to raise the pH. Table 13.1 may help to explain the relationship between pH and some common materials.

Table 13.1

H^+ concentration	Soil reaction	pH	Example substance
10^{-0}	pure acids	0	sulphuric acid
10^{-1}	extremely acid	1	battery acid, gastric juices
10^{-2}	extremely acid	2	worst acid rain record
10^{-3}	very strongly acid	3	cola, lemon, vinegar, sweat
10^{-4}	strongly acid	4	orange, lemonade
10^{-5}	moderately acid	5	urine
10^{-6}	slightly acid	6	clean rain, coffee
10^{-7}	**neutral**	**7**	**distilled water**
10^{-8}	slightly alkaline	8	sea water, baking soda
10^{-9}	moderately alkaline	9	soap solution
10^{-10}	strongly alkaline	10	milk of magnesia
10^{-11}	very strongly alkaline	11	strong bleach
10^{-12}	extremely alkaline	12	ammonia smelling salts
10^{-13}	extremely alkaline	13	caustic soda solution
10^{-14}	pure alkaline	14	concentrated sodium hydroxide

pH range and plant tolerance

The following list gives the pH tolerance range for the most common amenity and commercial crop production species. The optimum pH would normally be in the middle of the range. Where only a lower figure is given, this is the threshold below which plant growth is known to suffer.

Table 13.2

Name	pH range	Name	pH range	Name	pH range
Loam and mineral soil mediums					
Nursery stock (including ornamental trees, flowers and shrubs)					
Abelia	6.0–8.0	Acers	5.5–6.5	Ajuga	4.0–6.0
Acacia	6.0–8.0	Adonis	6.0–8.0	Althea	6.0–7.5
Acanthus	6.0–7.0	Ageratum	6.0–7.5	Alyssum	6.0–7.5
Aconitum	5.0–6.0	Ailanthus	6.0–7.5	Amaranthus	6.0–7.5

(Continued)

Table 13.2 (Continued)

Name	pH range	Name	pH range	Name	pH range
Nursery stock (including ornamental trees, flowers and shrubs) – continued					
Anchusa	6.0–7.5	Chrysanthemum	6.0–7.0	Gazania	5.5–7.0
Androsace	5.0–6.0	Cissus	6.0–7.5	Gentiana	5.0–7.5
Anemone	6.0–7.5	Cistus	6.0–7.5	Geum	6.0–7.5
Anthyllis	5.0–6.0	Clarkia	6.0–6.5	Ginko	5.5–7.0
Arbutus	4.0–6.0	Cleanthus	6.0–7.5	Gladioli	6.0–7.0
Ardisia	6.0–8.0	Clematis	5.5–7.0	Globularia	5.5–7.0
Arenaria	6.0–8.0	Colchicum	5.5–6.5	Godetia	6.0–7.5
Aristia	6.0–7.5	Columbine	6.0–7.0	Goldenrod	5.0–7.0
Armeris	6.0–7.5	Convolvulus	6.0–8.0	Gourd	6.0–7.0
Arnice	5.0–6.5	Coreopsis	5.0–6.0	Gypsophilia	6.0–7.5
Asperula	6.0–8.0	Coronilla	6.5–7.5	Hawthorn	6.0–7.0
Ashodoline	6.0–8.0	Corydalis	6.0–8.0	Helianthus	5.0–7.0
Aster	5.5–7.5	Cosmos	5.0–8.0	Helleborus	6.0–7.5
Astilbe	6.0–8.0	Cotoneaster	6.0–8.0	Hibiscus	6.0–8.0
Aubretia	6.0–7.5	Crab apple	6.0–7.5	Holly	5.0–6.5
Avens	6.0–7.5	Crocus	6.0–8.0	Holyhock	6.0–8.0
Azalea	4.5–6.0	Cynoglossum	6.0–7.5	Horse chestnut	5.5–7.0
Beauty bush	6.0–7.5	Daffodil	6.0–6.5	Hydrangeas	
Bergenia	6.0–7.5	Dahlia	6.0–7.5	blue	4.0–5.0
Betula papyrifera	5.5–6.5	Day lily	6.0–8.0	pink	6.0–7.0
Bleeding heart	6.0–7.5	Delphinium	6.0–7.5	white	6.5–8.0
Bluebell	6.0–7.5	Deutzia	6.0–7.5	Hypericum	5.5–7.0
Broom	5.0–6.0	Dianthus	6.0–7.5	Iris	5.0–6.5
Buddleia	6.0–7.0	Dogwood	5.0–6.5	Ivy	6.0–8.0
Bupthalum	6.0–8.0	Elaeagnus	5.0–7.5	Juniper	5.0–6.5
Calendula	5.5–7.0	Enkianthus	5.0–6.0	Kalmia	4.5–5.0
Calluna vulgaris	4.5–5.5	Erica carnea	4.5–5.5	Kerria	6.0–7.0
Camassia	6.0–8.0	Erica cinerea	5.5–6.5	Laburnum	6.0–7.0
Camellia	4.5–6.0	Eucalyptus	4.0–6.5	Laurel	4.5–6.0
Candytuft	6.0–7.5	Euphorbia	6.0–7.0	Larch	4.5–7.5
Canna	6.0–8.0	Everlasting	5.0–6.0	Lavender	6.5–7.5
Canterburybells	6.0–7.5	Fir	4.0–6.5	Liatris	5.5–7.5
Carnation	6.0–7.5	Firethorn	6.0–8.0	Ligustrum	5.0–7.5
Catalpa	6.0–8.0	Forget-me-not	6.0–8.0	Ligustrum	
Celosia	6.0–7.0	Forsythia	6.0–8.0	ovali folium	6.5–7.0
Centaurea	5.0–6.5	Foxglove	6.0–7.5	Lilac	6.0–7.5
Cerastium	6.0–7.0	Fritillaria	6.0–7.5	Lily-of-the-valley	4.5–6.0
Chamaecyparis		Fuchsia	5.5–6.5	Lithospermum	5.0–6.5
lawsoniana	4.5–5.5	Gaillardia	6.0–7.5	Lobelia	6.0–7.5
'Columnaris'		Gardenia	5.0–7.0	Lupinus	5.5–7.0

(Continued)

Table 13.2 (Continued)

Name	pH range	Name	pH range	Name	pH range
Nursery stock (including ornamental trees, flowers and shrubs) – continued					
Magnolia	5.0–6.0	Pelargonium	6.0–7.5	Sorbus aucuparia	6.5–7.0
Mahonia	6.0–7.0	Polyanthus	6.0–7.5	Soapwort	6.0–7.5
Marguerite	6.0–7.5	Poplar	5.5–7.5	Speedwell	5.5–6.5
Marigold	5.5–7.0	Poppy	6.0–7.5	Spirea	6.0–7.5
Molinia	4.0–5.0	Portulaca	5.5–7.5	Spruce	4.0–5.9
Moraea	5.5–6.5	Primrose	5.5–6.5	Stock	6.0–7.5
Morning glory	6.0–7.5	Privet	6.0–7.0	Stonecrop	6.5–7.5
Moss	6.0–8.0	Prunella	6.0–7.5	Sumach	5.0–6.5
Moss, sphagnum	3.5–5.0	Prunus	6.5–7.5	Sunflower	6.0–7.5
Mulberry	6.0–7.5	Pyracantha	5.0–6.0	Sweet william	6.0–7.5
Myosotis	6.0–7.0	Quercus (oak)	5.5–6.5	Sycamore	5.5–7.5
Narcissus	6.0–7.5	Red hot poker	6.0–7.5	Syringa	6.0–8.0
Nasturtium	5.5–7.5	Rhododendron	4.5–6.0	Syringa vulgaris	6.5–7.0
Nicotiana	5.5–6.5	Roses: hybrid tea	5.5–7.0	Tamarix	6.5–8.0
Pachysandra	5.0–8.0	climbing	6.0–7.0	Tobacco	5.5–7.5
Paeonia	6.0–7.5	rambling	5.5–7.0	Trillium	5.0–6.5
Pansy	5.5–7.0	Rosa laxa	6.5–7.0	Tulip	6.0–7.0
Passion flower	6.0–8.0	Rosa multiflora	5.5–6.5	Viburnum	5.0–7.5
Pasque flower	5.0–6.0	Rowan	4.5–6.5	Viola	5.5–6.5
Paulownia	6.0–8.0	Salix	6.5–7.0	Violet	5.0–7.5
Picea pungens	4.5–5.5	Salvia	6.0–7.5	Virginia creeper	5.0–7.5
Pea, sweet	6.0–7.5	Saintpaulia	6.0–7.0	Wallflower	5.5–7.5
Penstemon	5.5–7.0	Scabiosa	5.0–7.5	Water lily	5.5–6.5
Peony	6.0–7.5	Sea grape	5.0–6.5	Weyela	6.0–7.0
Periwinkle	6.0–7.5	Sedum	6.0–8.0	Wistine	6.0–8.0
Pittosporum	5.5–6.5	Snapdragon	5.5–7.0	Yew	5.0–7.5
Plantain	6.0–7.5	Snowdrop	6.0–8.0	Zinnia	5.5–7.5
Grasses					
Annual meadow grass	5.5–7.5	Canada	5.7–7.2	Pampas	6.0–8.0
Bents:	5.5–6.5	Crested dogstail	5.0–6.5	Rye	5.8–7.4
Browntop	4.5–6.5	Clover	6.0–7.0	Smooth stalked meadow	6.0–8.0
Creeping	5.5–7.5	Cock'sfooot	5.3	grass	
Colonial	5.6–7.0	Fescues:	6.0–7.5	Timothy	5.3
Velvet	5.2–6.5	Red	4.5–8.0	Trefoil	6.1
Bermuda grass	6.0–7.0	Chewing	4.5–8.0	Vetches	5.9
Bluegrass:		Hard	4.0–5.5	Yorkshire fog	4.6
Annual	5.5–7.0	Sheep's	4.0–5.5	Wheat grass	6.1–8.6
Kentucky	5.8–7.5	Tall	5.5–7.0		
Rough	5.8–7.2	Meadow	6.0–7.5		

(Continued)

Table 13.2 (Continued)

Name	pH range	Name	pH range	Name	pH range
Common weeds (wild plants) (Threshold below which plant growth is known to suffer)					
Annual meadow grass	5.5–7.5	Cowslip	5.1	Oxeye daisy	6.1
Bird's foot trefoil	6.1	Dog violet	5.1	Rest harrow	6.1
Burnuct	6.1	Gentian	6.1	Wild carrot	5.6
Cleavers	6.1	Kidney vetch	6.1	Wild clematis	6.1
Colt's foot	5.6	Milfoil	4.6	Woodrush	4.6
Common nettle	5.1	Milkwort	6.1	Yorkshire fog	4.6
Creeping softgrass	4.6				
Fruit crops					
Apple	5.0–6.5	white	6.0–8.0	Nectarine	6.0–7.5
Apricot	6.0–7.0	Damson	6.0–7.5	Papaw	6.0–7.5
Avocado	6.0–7.5	Gooseberry	5.0–6.5	Pear	5.3
Banana	5.0–7.0	Grapevine	6.0–7.5	Peach	6.0–7.5
Blackberry	4.9–6.0	Hazelnut	6.0–7.0	Pineapple	5.0–6.0
Blueberry	4.5–6.0	Hop	6.0–7.5	Plum	5.6–7.5
Cherry	6.0–7.5	Lemon	6.0–7.0	Pomegranate	5.5–6.5
Cranberry	4.0–5.5	Lychee	6.0–7.0	Quince	6.0–7.5
Currants:		Mango	5.0–6.0	Rasberry	5.5–6.5
black	6.0–8.0	Melon	5.5–6.5	Rhubarb	5.5–7.0
red	5.5–7.0	Mulberry	6.0–7.5	Strawberry	5.1–7.5
Vegetables, herbs and some arable crops					
Artichoke	6.5–7.5	Cauliflower	5.5–7.5	Lettuce	6.1–7.0
Asparagus	5.9–8.0	Celery	6.3–7.0	Linseed	5.4
Barley	5.9	Chicory	5.1–6.5	Marjoram	6.0–8.0
Basil	5.5–6.5	Corn, sweet	5.5–7.5	Marrow	6.0–7.5
Beans:		Cress	6.0–7.0	Millet	6.0–6.5
runner	6.0–7.5	Cotton	5.0–6.0	Mint	6.6–8.0
broad	6.0–7.5	Courgettes	5.5–7.0	Mushroom	6.5–7.5
french	6.0–7.5	Cowpea	5.0–6.5	Mustard	5.4–7.5
Beet, sugar	5.9	Cucumber	5.5–7.0	Oats	5.4
Beet, table	5.9	Fennel	5.0–6.0	Olive	5.5–6.5
Beetroot	6.0–7.5	Garlic	5.5–7.5	Onion	5.7–7.0
Broccoli	6.0–7.0	Ginger	6.0–8.0	Paprika	7.0–8.5
Brussel sprouts	5.7–7.5	Ground nut	5.5–6.5	Parsley	5.1–7.0
Cabbage	5.4–7.5	Horseradish	6.0–7.0	Parsnip	5.4–7.5
chinese	6.0–7.5	Kale	5.4–7.5	Pea	5.9–7.5
Calabrese	6.5–7.5	Kohlrabi	6.0–7.5	Peanut	5.0–6.5
Cantaloupe	6.0–8.0	Leek	5.8–8.0	Pepper	5.5–7.0
Carrots	5.7–7.0	Lentil	5.5–7.0	Peppermint	6.0–7.5

(Continued)

Table 13.2 (Continued)

Name	pH range	Name	pH range	Name	pH range
Pistacio	5.0–6.0	Rhubarb	5.4	Sunflower	6.0–7.5
Potato	4.9–6.0	Sage	5.5–6.5	Swede	5.4–7.5
Potato, sweet	5.5–6.0	Shallot	5.5–7.0	Thyme	5.5–7.0
Pumpkin	5.5–7.5	Sorghum	5.5–7.5	Tomato	5.1–7.5
Radish	6.0–7.0	Soya bean	5.5–6.5	Turnip	5.4–7.0
Rape	5.6	Spearmint	5.8–7.5	Water cress	6.0–8.0
Rice	5.0–6.5	Spinach	6.0–7.5	Water melon	5.0–6.5
Rosemary	5.0–6.0	Sugarcane	6.0–8.0	Wheat	5.5–7.5

Loamless compost

Container-grown nursery stock and protected pot plants (house plants)

Name	pH range	Name	pH range	Name	pH range
General plants	5.5–6.5	Campanula	5.5–6.5	Dutchman's pipe	6.0–8.0
Abutilon	5.5–6.5	Capscium pepper	5.0–6.5	Easter lily	6.0–7.0
Acorus	5.0–6.5	Cardinal flower	5.0–6.0	Ericaceous	5.0–5.5
Aechmea	5.0–5.5	Carnations	5.5–6.5	Elephant's ear	6.0–7.5
African violet	6.0–7.0	Castor oil plant	5.5–6.5	Episcia	6.0–7.0
Aglaonema	5.0–6.0	Century plant	5.0–6.5	Eucalyptus	6.0–8.0
Amaryllis	5.5–6.5	Chinese primrose	6.0–7.5	Euonymus	5.5–7.0
Anthurium	5.0–6.0	Christmas cactus	5.0–6.5	Feijoa	5.0–7.5
Aphelandra	5.0–6.0	Chrysanthemums	5.5–6.5	Ferns:	
Aralia	6.0–7.5	Cineraria	5.5–7.0	bird's nest	5.0–5.5
Araucaria	5.0–6.0	Clerodendrum	5.0–6.0	boston	5.5–6.5
Asparagus fern	6.0–8.0	Clivia	5.5–6.5	christmas	6.0–7.5
Aspen	4.0–5.5	Cockscomb	6.0–7.5	cloak	6.0–7.5
Aspidistra	4.0–5.5	Coffee plant	5.0–6.0	feather	5.5–7.5
Azalea	4.5–6.0	Coleus	6.0–7.0	hart's tonge	7.0–8.0
Bedding plants	5.5–6.5	Columnea	4.5–5.5	holly	4.5–6.0
Begonia	5.5–7.0	Coral berry	5.5–7.5	maidenhair	6.0–8.0
Bird of paradise	6.0–6.5	Crassula	5.0–6.0	rabbits foot	6.0–7.5
Bishop's cap	5.0–6.0	Creeping fig	5.0–6.0	spleenwort	6.0–7.5
Black-eyed Susan	5.5–7.5	Croton	5.0–6.0	Fig	5.0–6.0
Blood leaf	5.5–6.5	Crown of thorns	6.0–7.5	Fittonia	5.5–6.5
Bottle brush	6.0–7.5	Cuphea	6.0–7.5	Freesia	5.5–6.5
Bougainvillea	5.5–7.5	Cyclamen	6.0–7.0	French marigold	5.0–7.5
Bromeliads	5.0–6.0	Cyperus	5.0–7.5	Gardenia	5.0–6.0
Butterfly flower	6.0–7.5	Daphne	6.5–7.5	Genista	6.5–7.5
Cactus	4.5–6.0	Dieffenbachia	5.0–6.0	Geranium	6.0–8.0
Calceolaria	6.0–7.0	Dipladenia	6.0–7.5	Gloxinia	5.5–6.5
Caladium	6.0–7.5	Dizygotheca	6.0–7.5	Grape hyacinth	6.0–7.5
Calla lily	6.0–7.0	Dracaena	5.0–6.0	Grape ivy	5.0–6.5
Camellia	4.5–5.5	Dragon tree	5.0–7.5	Grevillea	5.5–6.5

(Continued)

Table 13.2 (Continued)

Name	pH range	Name	pH range	Name	pH range
Container-grown nursery stock and protected pot plants (house plants) (continued)					
Gynura5.5	5.5–6.5	Monstera	5.0–6.0	Rubber plant	5.0–6.0
Hedera (ivy)	6.0–7.5	Myrtle	6.0–8.0	Sansevieria	4.5–7.0
Heliotropium	6.0–8.0	Nephthytis	4.5–5.5	Saxifraga	6.0–8.0
Helxine	5.0–6.0	Never never plant	5.0–6.0	Schizanthus0	6.0–7.0
Herringbone	5.0–6.0	Nicodernia (indoor oak)	6.0–8.0	Scilla	6.0–8.0
Hibiscus	6.0–8.0	Oleander	6.0–7.5	Scindapsus	5.0–6.0
Hyacinth	6.5–7.5	Oplisemenus	5.0–6.0	Seed compost	5.5–6.5
Hoya	5.0–6.5	Orange plant	6.0–7.5	Selaginella	6.0–7.0
Impatiens	5.5–6.5	Orchid	4.5–5.5	Senecio	6.0–7.0
Indegofera	6.0–7.5	Oxalis	6.0–8.0	Shrimp plant	5.5–6.5
Iresine	5.0–6.5	Painted lady	6.0–7.5	Spanish bayonet	5.5–7.0
Ivy tree	6.0–7.0	Palms	6.0–7.5	Spider plant	6.0–7.5
Jacaranda	6.0–7.5	Pandanus	5.0–6.0	Succulents	5.0–6.5
Japanese sedge	6.0–8.0	Patient lucy	5.5–6.5	Syngunium	5.0–6.0
Jasminum	5.5–7.0	Peacock plant	5.0–6.0	Thunbergia	5.5–7.5
Jerusalem cherry	5.5–6.5	Pelagoniums	6.0–7.5	Tolmiea	5.0–6.0
Jessamine	5.0–6.0	Pellionia	5.0–6.0	Tomato	5.5–6.0
Kaffir	6.0–7.5	Philodendron	5.0–6.0	Tradescanthia	5.0–6.0
Kalanchoe	6.0–7.5	Phlox	5.0–6.5	Umbrella tree	5.0–7.5
Kangaroo thorn	6.0–8.0	Pitcherplant	4.0–5.5	Venus flytrap	4.0–5.0
Kangaroo vine	5.0–6.5	Pilea	6.0–8.0	Verbena	6.0–8.0
Lace flower	6.0–7.5	Plumbago	5.5–6.5	Vinca	6.0–7.5
Lantana	5.5–7.0	Podacarpus	5.0–6.5	Weeping fig	5.0–6.0
Laurus bay tree	5.0–6.0	Poinsettia	6.0–7.5	Yucca	6.0–8.0
Lemon plant	6.0–7.5	Polyscias	6.0–7.5	Zebrina	5.0–6.0
Mimosa	5.0–7.0	Pothos	5.0–6.0		
Mind your own business	5.0–5.5	Prayer plant	5.0–6.0		
		Puncia	5.5–6.5		

Exercise 13.1

Soil and compost pH testing

Background

Before we can modify growth medium pH it is necessary to determine the existing level of soil acidity. Comparison can then be made between this and the plant pH tolerance range for the species growing. If the acidity level is too low, it will need to be corrected by liming. If it is too high, it will need to be lowered, normally by applying sulphur.

There are several methods available to determine pH levels in the growth medium. These range from litmus indicator paper, garden centre type needle probe (e.g. Rapitest), test tube measurements (e.g. BDH method), portable electrode meters (e.g. Hanna Instruments pHep range) and laboratory based analysis (e.g. ADAS or Defra). In laboratory tests pH is measured by using a hydrogen electrode in the solution of interest as one half of a cell, and a reference electrode (e.g. a calomel electrode) as the other half cell. Often these are combined into a single pH electrode that lasts only about six months before needing replacement.

Of these, the test tube method is most suitable for determining quickly, on site, the pH of soils and composts, although the BDH method will be considered. Several products are available that use similar principles including Rapitest and Sudbury products, ADAS field kits and many amateur kits are available from garden centres.

The BDH method for pH testing

British Drug House (BDH, now owned by Merkoquant) developed the apparatus for the following technique to determine soil pH. It involves mixing a small amount of soil with a white powder called barium sulphate, some distilled water and universal indicator solution. The cocktail is then shaken up and left to stand. The barium sulphate causes the soil to fall to the bottom of the test tube (flocculate). It is a neutral material, and like distilled water, does not influence the pH. After a few minutes the soil particles will settle out, leaving a coloured solution above. This coloured solution is then compared with a colour chart and the corresponding pH recorded.

Aim

To determine the pH of soil and compost growth mediums using the BDH method.

Apparatus

BDH test tubes	Barium sulphate
Universal indicator solution	Distilled water
Soil A	Soil B
Potting compost A	Ericaceous compost B
pH colour chart	

Useful websites

www.sciencepages.co.uk

www.avogadro.co.uk

Method

Follow the method in Figure 13.2 to find the pH of the different materials.

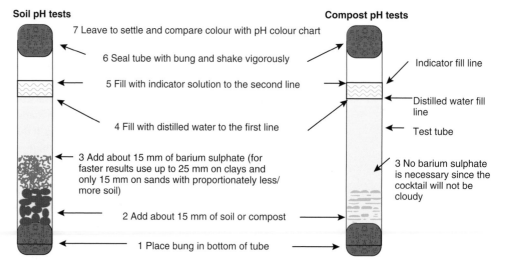

Soil pH tests   **Compost pH tests**

7 Leave to settle and compare colour with pH colour chart

6 Seal tube with bung and shake vigorously

Indicator fill line

5 Fill with indicator solution to the second line

Distilled water fill line

4 Fill with distilled water to the first line

Test tube

3 Add about 15 mm of barium sulphate (for faster results use up to 25 mm on clays and only 15 mm on sands with proportionately less/more soil)

3 No barium sulphate is necessary since the cocktail will not be cloudy

2 Add about 15 mm of soil or compost

1 Place bung in bottom of tube

Figure 13.2 Soil and compost pH tests

Results

Record your results in the table provided.

Growth medium	pH	Interpretation (e.g. slightly acid)
Soil A		
Soil B		
Compost A		
Compost B		

Conclusions

1. Explain the purpose of using barium sulphate.
2. State the ideal pH for most plants grown in a soil.
3. Comment on the suitability of Soil A for growing white-flowered hydrangeas.
4. How does the method vary, if testing a compost?
5. What is the ideal pH for most plants grown in a compost?
6. Comment on the suitability of Soil B for growing strawberries.
7. Comment on the suitability of Compost A for growing azaleas
8. Comment on the suitability of Compost B for growing busy lizzies (*Impatiens*).
9. Explain what is meant by a 'calcicole plant' and give one example.
10. Explain what is meant by a 'calcifuge plant' and give one example.

Exercise 13.2

Soil texture assessment

Background

In addition to determining the soil pH, information about the soil texture is also needed. Soil texture refers to the relative proportions of sand, silt and clay particles in the soil. This is important because clay soils require more lime to correct acidity than sandy soils. This property is called buffering capacity.

Please refer to Chapter 10 for a detailed procedure for assessing soil texture.

Aim

To correctly conduct a hand texture assessment of different soils types.

Apparatus

  Figure 10.7
  Sandy soil A
  Clayey soil B

Method

Assess the texture of Soil A and Soil B, giving the reasoning behind your answer.

Results

Record your results in the table provided in the next page.

Sample	Texture	Reasoning
Soil A		
Soil B		

Conclusions

1. Explain which soil has a greater buffering capacity and also therefore ability to hold nutrients.
2. Which soil will require more lime to correct any soil acidity problem, and why?
3. How would the procedure to measure pH, using the BDH method, vary between these two soils?

Answers

Exercise 13.1. Soil and compost pH testing

Results

Soil A should be selected to give a slightly alkaline reading.

Soil B should be selected to give an acid reading.

Potting compost A has a pH of 6.0.

Ericaceous compost B has a pH of 5.5.

Conclusions

1. Barium sulphate is used to flocculate clay particles. causing them to sink to the bottom of the test tube and leaving a clear liquid above which can be easily compared with the colour chart.

2. pH 6.5.

3. White flowered hydrangeas have a pH range of 6.5–8.0, so soil a at pH 7.0 would be suitable.

4. When testing a compost barium sulphate is not required since the cocktail will not be cloudy and the colour can be easily compared with the colour chart.

5. pH 5.8

6. Strawberries have a pH range of 5.1 to 7.5; if soil B is within this range it will be suitable, but will grow best as the mid point.

7. Azaleas are a calcifuge species having a pH range from 4.5 to 6.0. Compost A at pH 6.0 would not be suitable because it is at the edge of the tolerance range. A more acid compost should be selected.

8. Busy lizzies (*Impateins*) have a pH range from 5.5 to 6.5. compost B would not be suitable since it has a pH at the edge of the tolerance range. A less acid compost should be selected.

9. A calcicole plant is lime-tolerant, e.g. geranium (pH 6.0–8.0) and firethorn (pH 6.0–8.0).

10. A calcifuge plant is acid-loving, e.g. azalea (pH 4.5–6.0) and bird's nest fern (pH 5.0–5.5).

Exercise 13.2. Soil texture

Results

A sandy textural class should be selected for soil A and a clayey texture for soil B.

Conclusions

1. Clay has a greater buffering capacity because it has cation exchange capacity, giving it an ability to hold nutrients. It has a larger surface are per unit volume than sand and will therefore hold more water.

2. The clay will require more lime because it has a higher buffering capacity.

3. Clay would require more barium sulphate than sand to help flocculation.

Chapter 14   Raising soil pH

Key facts

1. Soil pH becomes more acidic over time due to rainfall, micro-organisms and root exudates and require liming.
2. Lime is any material used to correct soil acidity. Examples include ground limestone and magnesian limestone.
3. The effectiveness of liming materials is determined through their neutralizing values when compared to that of with calcium oxide (CaO).
4. A lime dressing is the quantity of lime required to raise the top 150 mm of soil to pH 6.5.
5. The quantity of lime required varies greatly with soil texture. Clays require much more than sandy soils.

BROWN   978 0 7506 8702 7

Background

Over a period of years the pH of the growth medium will naturally tend to become more acidic. This is due to a variety of processes. For example, rainfall washes out (leaches) free lime (calcium carbonate, $CaCO_3$) from the soil and in addition acid rainfall builds up the hydrogen ion concentration. Some plants, such as ericaceous and conifer species, acidify the soil due to organic acids being leached from their leaves. Micro-organism activity may vary pH by up to one pH unit over the course of a year, due to the gases given off during respiration and other chemical reactions.

The signs of soil acidity problems in plants include: loss of vigour, leaf discoloration, reduced fruit, stunted roots and the growth of weeds. Low pH also prevents some nutrients from being available to plants including phosphorus, potassium, magnesium and calcium. Other physiological damage such as club-root in brassicas may be seen. An acid pH hinders soil organisms from nitrogen fixing and recycling and soil structure will be damaged. In acid soils of pH 5 or below, aluminium and manganese are released in toxic amounts to plants.

Once a soil acidity problem has been identified, action will need to be taken to correct it. Any material used to correct soil acidity is known as **lime**. This must be applied to neutralize both active and reserve hydrogen ions. The amount of lime needed will be determined by reference to:

- present pH
- soil texture and buffering capacity
- effectiveness of the liming material (neutralizing value).

Exercise 14.1

Soil texture and lime requirements

Background

Soil texture greatly influences the amount of lime that is required to raise pH. It is more difficult to alter the pH of a clay soil than of a sandy soil due to different buffering capacities.

The amount of lime required to change pH is given in Table 14.1. By convention this is always stated using ground limestone, in tonnes per hectare. This can

Table 14.1 Lime required to raise pH to 6.5, in tonnes/hectare of ground limestone (multiply by 100 for g/m^2)

Present pH	Loamy sands	Sandy loams	Silt loams	Loams	Clay loams
3.5	15.0	18.1	21.0	24.0	31.0
4.0	12.5	15.0	17.5	20.0	25.0
4.5	10.0	11.9	14.0	16.2	20.0
5.0	7.5	9.4	10.0	11.9	15.0
5.5	5.0	6.2	7.0	8.1	10.0
6.0	2.5	3.1	3.5	3.7	5.0

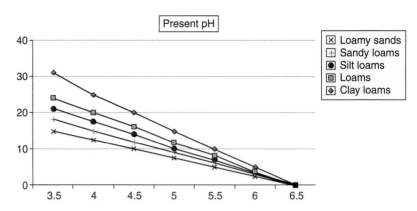

Figure 14.1 Graph using data from Table 14.1

be converted to grammes per metre square, by multiplying by 100. Figure 14.1 shows this data plotted into a graph for future reference.

Aim

To gain competence at gathering and interpreting information on soil texture and liming requirements to correct soil acidity.

Apparatus

   Table 14.1
   Example situations

Useful websites

www.aglime.org.uk

www.rhs.org.uk

Method

Either using Table 14.1, or a graph plotted from the data (see Figure 14.1), state the liming requirement to raise pH to pH 6.5 under the conditions given below.

Results

Enter your results in the table provided.

State the amount of lime required to raise the pH of	Lime requirement (ground limestone) t/ha	Lime requirement (ground limestone) g/m^2
1. A clay loam cricket square from pH 4.5 to pH 6.5		
2. A clay bedding plant area from pH 5.0 to pH 6.5		
3. A loamy sand old lawn from pH 5.5 to pH 6.5		
4. A rugby pitch (silt loam) from pH 4.5 to pH 6.5		
5. Allotments (loams) from pH 5.0 to pH 6.5		
6. An unused acid bed (sandy loam) from pH 4.0 to pH 6.5		

Conclusions

1. Explain how the pH of soil would naturally become more acidic.
2. Which sites have the greatest buffering capacity?
3. What is the effect of an inappropriate level of soil acidity on plant growth?
4. Explain what is meant by the term 'lime'.

Liming materials and neutralizing values

Background

Any material used to correct soil acidity is called lime.

Lime must dislodge hydrogen and aluminium ions from receptor molecules and then neutralize them in the soil solution. The calcium carbonate in lime reacts with hydrogen ions to form both carbonic acid and calcium ions in the soil solution. The carbonic acid can change into carbon dioxide, escaping into the atmosphere or dissolving in soil water.

A lime requirement is the amount of lime required to raise the top 150 mm of the soil to pH 6.5. For the best results finely ground lime should be applied several months before planting. The application of liming materials generally corrects soil acidity by neutralizing and replacing hydrogen ions with calcium ions.

There are many different forms of lime, which all have different calcium ion contents. Because of this, they will vary in their effectiveness to correct soil acidity. Limes with a high calcium ion content will be more effective than limes with a low calcium ion content and will therefore require less material to neutralize acidity.

We can compare the effectiveness of different liming materials by referring to their neutralizing values (NV). All liming materials are compared with pure calcium oxide (CaO). The neutralizing value compares how effective the liming material is, against using pure calcium oxide. For example, if using ground limestone, twice as much material would be needed than if using pure calcium oxide. Ground limestone therefore has a neutralizing value of 50, or is only 50 per cent as effective in correcting soil acidity as pure calcium oxide. Pure calcium oxide has a neutralizing value of 100. The neutralizing values may vary slightly between batches, but will always be stated on the bag that the lime is sold in. A list of some common liming materials and their neutralizing values is given in Table 14.2.

A lime requirement is always stated in terms of ground limestone (in t/ha). However, it may be more appropriate to use a cheaper or more convenient alternative liming material and by comparing their neutralizing values an equivalent application rate can be found.

If we know the quantity of ground limestone required to raise the pH to 6.5, and we know the neutralizing value of the alternative material, we can calculate how much of the alternative liming material is needed, using the following formulae:

$$\text{Lime requirement of alternative material (t/ha)} = \text{Quantity of ground limestone recommended} \times \frac{\text{NV ground limestone}}{\text{NV alternative material}}$$

Table 14.2 Common liming materials

Material	Neutralizing value
burnt lime (quicklime, calcium oxide)	85–90
burnt magnesian limestone	95–110
calcareous shell sand	24–45
dolomitic limestone	48
ground chalk	48–50
ground limestone	48–50
ground magnesian limestone	50–55
hydrated lime (slaked lime)	70
marl	variable
Pure calcium carbonate	56
pure calcium hydroxide	74
pure calcium oxide	**100**
screened chalk	45
waste limes (basic slag)	variable

For example, an application rate of 2.5 t/ha ground limestone (NV50), is recommended to correct soil acidity on a loamy sand land reclamation site. It is more convenient to use a local suppler of screened chalk (NV45). Following the above equation the conversion is as follows:

$$\text{Amount of screened chalk (NV45) required} = \frac{2.5 \text{ t/ha ground limestone}}{1} \times \frac{50 \text{ (NV ground limestone)}}{45 \text{ (NV screened chalk)}}$$

$$= 2.5 \times \frac{50}{45}$$

$$= 2.78 \text{ t/ha screened chalk}$$

Aim

To develop skills at calculating the quantity of alternative liming materials required to correct soil acidity and raise pH to 6.5.

Apparatus

    Table 14.1
    Table 14.2
    Sample situations given below

Useful website

www.aglime.org.uk

Method

Using the table of liming materials and the formulae provided, calculate an alternative liming requirement for the situations given below.

Results

1. An application rate of 10 t/ha ground limestone (NV50) is recommended to raise the pH from 5.5 to 6.5 on the amenity beds (clay loam) behind a housing development. What is the alternative requirement for:
   (a) calcareous shell sand (NV25)?
   (b) hydrated lime (NV 70)?

2. An application rate of 14 t/ha ground limestone (NV50) is recommended to raise the pH from 4.5 to 6.5 on an unused acid bed (silt loam texture). What is the alternative requirement for:
   (a) calcareous shell sand (NV25)?
   (b) screened chalk (NV 45)?
3. An application rate of 2.5 t/ha ground limestone (NV50) is recommended to raise the pH from 6.0 to 6.5 on a utility lawn (loamy sand texture). What is the alternative requirement for:
   (a) calcareous shell sand (NV25)?
   (b) pure calcium oxide (NV 100)?

Conclusions

1. Which liming material is most commonly available in garden centres?
2. What information about the liming material needs to be recorded from every packet of lime?
3. How are t/ha converted to g/m2?
4. When should lime 'ideally' be applied?

Exercise 14.3

Integration of skills

Background

Sample soil A and sample soil B are both taken from recreational areas in different parts of the county. The growth of plants has been very poor and this is thought to be due to an acid soil pH. Determine whether or not this is true, and if so, produce a liming recommendation to correct the soil acidity to pH 6.5. However, as an additional consideration, the County Council Parks Department has stated that they only stock calcareous shell sand (NV25), screened chalk (NV45), and hydrated lime (NV70). Your recommendations must therefore be for these liming materials.

Aim

To use all the skills developed in these exercises to calculate a lime requirement for soil A and soil B.

Method

Investigate the soil samples and complete the table provided.

Results

Enter your results in the table provided.

Soil sample	Example soil	Soil A	Soil B
Present pH	4.5		
Texture	loamy sand		
Lime requirement (ground limestone, NV50) (t/ha)	10.0		
Conversion to use calcareous shell sand (NV 25) (t/ha)	$10 \times 50/25 = 20$		
Conversion to use screened chalk (NV 45) (t/ha)	$10 \times 50/45 = 11$		
Conversion to use hydrated lime (NV70) (t/ha)	$10 \times 50/70 = 7.14$		

Conclusions

1. Explain what is meant by the term 'lime'.
2. List six different liming materials.
3. How would you convert t/ha to g/m^2?

Answers

Exercise 14.1. Soil texture and lime requirement

Results

1.   20 t/ha, 2000 g/m^2

2.   15 t/ha, 1500 g/m^2

3.   5 t/ha, 500 g/m^2

4.   14 t/ha, 1400 g/m^2

5.   11.9 t/ha, g/m^2

6.   15 t/ha, 1500 g/m^2

Conclusions

1.   The soil naturally becomes acidic due to: the activity of microorganisms, acid rain, and plant acid exudes.

2.   Sites with the most clay have the greatest buffering capacity: sites 2 and 1.

3.   If the soil is too acidic plant growth will be inhibited. Many nutrients become 'locked out' and unavailable to the plant.

4.   Lime is any material used to correct soil acidity.

Exercise 14.2.   Liming materials and neutralizing values

Results

1.   (a) $10 \times 50 / 25 = 20$ t/ha
       (b) $10 \times 50 / 70 = 7.14$ t/ha

2.   (a) $14 \times 50 / 25 = 28$ t/ha
       (b) $14 \times 50 / 45 = 15.55$ t/ha

3.   (a) $2.5 \times 50 / 25 = 5$ t/ha
       (b) $2.5 \times 50 / 100 = 1.25$ t/ha

Conclusions

1.   Ground limestone (NV 55).

2.   The neutralizing value (NV).

3.   Multiply by 100.

4.   Finely ground several months before planting

Exercise 14.3.   Integration of skills

Results

First determine the texture and the present pH. Then use Table 14.1 to determine the quantity of ground limestone required. Next use the formulae from Exercise 14.2 to calculate the quantities of alternative liming materials needed.

Conclusions

1.   Lime is any material used to correct soil acidity.

2.   Any six materials from Table 14.2.

3.   Multiply by one hundred.

Chapter 15   Lowering soil pH (increasing soil acidity)

Key facts

1. Soils that are too limey are called alkaline. These conditions can arise through poor management.
2. Alkaline conditions damage soil structure and result in poor growth.
3. Sulphur-based products are used to lower pH which are changed into sulphuric acid in the soil.
4. The free-lime (calcium carbonate) content of the soil must be determined using 10 per cent hydrochloric acid, and then neutralized using ground sulphur.
5. The sulphur required to lower pH varies with soil texture which must also be assessed.

Background

Soils that are too alkaline will require special corrective treatments and good management to achieve healthy plant growth. Problems of soil alkalinity normally occur due to an increasing proportion of sodium ions present in the soil solution. This may occur for a variety of reasons such as:

- rising water tables redepositing previously leached ions (e.g. near flooding rivers)
- high dissolved nutrient content in irrigation water
- poor drainage, preventing leaching of dissolved ions
- poor watering practice, such as excessive drying out between irrigations.

Alkaline conditions cause the separation of soil clay particles and the breakdown of organic matter. Consequently, these soils may also have structural problems such as cracking in hot weather and becoming puddled when wet. Nitrogen and calcium deficiency are also common symptoms.

In other situations it may simply be that the intended plants pH tolerance range is lower than the present soil pH and measures to lower pH (increase acidity) will be required (e.g. lowering pH of turfed areas to pH 6.0 to discourage weed growth).

Lowering the pH of the growth medium is a difficult and expensive processes. It is normally achieved by the application of acid-forming sulphur substances, such as ground sulphur (rolls of sulphur), ferrous sulphate, aluminium sulphate, low-grade pyrites and calcium sulphate (gypsum). The sulphur is changed to sulphuric acid by the action of soil micro-organisms and weathering processes. It can therefore cause plant death if applied to sites containing growing plants. Ideally it should be applied several months before planting. There are also several commercial products available that claim to lower the soil's pH over a period of months, for example Phostrogen and Liquid Sod (flowerable sulphur) products.

As a cheaper alternative acid-based fertilizers can also help where only a slight adjustment is desired, possibly over a period of years. Likewise, sphagnum moss peat is naturally acidic (about pH 5.0) and this can be incorporated around the rootball for the rapid establishment of the plant (e.g. when growing ericaceous heathers in chalky soils).

The amount of sulphur needed to increase acidity is based on a number of existing soil properties. These are:

- existing soil pH
- soil texture
- free lime (calcium carbonate content).

The practice of measuring pH and assessing texture have been covered in Chapters 13 and 14, and the following exercises on these characteristics are offered for wholeness and practice.

Soil and compost pH testing

Aim

To assess soil and compost pH using the BDH method.

Apparatus

2 alkaline soils labelled sample soil A and B
Compost A (sphagnum moss peat)
Compost B (ericaceous compost)
Compost C (coir compost)
BDH pH testing kit

Method

Assess the pH of sample soils A and B, and sample composts A, B and C.

Results

Record your results in the table provided.

Growth medium	pH	Interpretation (e.g. slightly alkaline)
Sample soil A		
Sample soil B		
Compost A		
Compost B		
Compost C		

Conclusions

1. State which soil has the greatest alkalinity problem.
2. Explain how the alkalinity problem may have developed.
3. Explain how compost may be used to help rhododendrons establish on a slightly acid soil.

Soil texture

Aim

To correctly conduct a hand texture assessment of sample soil A and sample soil B for the purpose of providing a subsequent recommendation to lower pH.

Apparatus

alkaline sample soil A and sample soil B
Figure 10.7
Table 13.2

Method

Assess the texture of sample soil A and sample soil B using the procedure for hand texture testing previously introduced in Chapter 10 (see Figure 10.7).

Results

Record your results in the table provided in the next page.

Sample	Texture
Sample soil A	
Sample soil B	

Conclusions

1. Which soil has a greater buffering capacity?
2. Which soil will require more sulphur to lower the pH and why?
3. List three plant species that are suitable to be grown in these soils without altering the existing pH.

Exercise 15.3

Free lime (calcium carbonate) content

Background

Before we can effectively lower pH we must assess the free lime or calcium carbonate ($CaCO_3$) content of the soil and neutralize it through the addition of ground sulphur. This may not lower soil pH, merely neutralize the free lime. Once this has been done we can then apply additional sulphur to lower pH units further.

The free lime content is assessed by dropping a weak acid solution (10 per cent hydrochloric) onto the soil. The acid will neutralize some of the free lime and the reaction is observed by watching the size of the bubbles and listening to the sound of the fizz (hold close to ear). Although the industry standard is to use 10 per cent hydrochloric acid alternative materials such as vinegar may be used. For each 0.1 per cent of free lime in the soil, 1 t/ha ground sulphur will be required to neutralize it. Table 15.1 states the quantities required and lists the reactions.

Aim

To assess the free lime ($CaCO_3$) content and sulphur neutralizing requirement of a range of horticultural materials.

Table 15.1 Free lime content ($CaCO_3$) and neutralizing sulphur requirement

Description of free lime status	$CaCO_3$ content %	Reaction on contact with 10% hydrochloric acid		Ground sulphur required to neutralize free lime content (t/ha)
		Visible effects	Audible effects	
non-calcareous	0.1	none	none	1
very slightly calcareous	0.5	none	slightly audible	5
slightly calcareous	1.0	slight fizz on individual grains just visible	faintly audible	10
slightly calcareous	2.0	slightly more general fizz on close inspection	moderate – distinctly audible	20
calcareous	5.0	moderate fizz, bubbles up to 3 mm diameter	easily audible	50
very calcareous	10.0	general strong fizz, large bubbles up to 7 mm diameter easily seen	easily audible	100

Apparatus

Sample soil A and sample soil B
5 types of sand and grit
Petri dishes
Vinegar
10 per cent hydrochloric acid solution (Warning – irritant and corrosive)

Method

1. Place a small amount of sample material onto a petri dish.
2. Slowly add a few drops of 10 per cent hydrochloric acid (warning – corrosive and irritant) onto the material and observe the reaction in accordance with Table 15.1.
3. Replicate the experiment, this time using vinegar.
4. State the free lime content of the material measured by each acid.
5. Record the amount of sulphur required to neutralize the free lime content of the materials.

Results

Record your results in the table provided.

Horticultural material	Free lime (CaCO$_3$ content)		Ground sulphur required to neutralize free lime (t/ha)
	10% hydrochloric acid	vinegar	
sea shells			
beach sand			
grit			
silver sand			
horticultural sand			
sample soil A			
sample soil B			

Conclusion

1. What can you conclude about the use of vinegar as a substitute for 10 per cent hydrochloric acid?
2. Explain why sea shells are unsuitable for lowering soil pH.
3. Which sample soil has the greatest free lime content?
4. Explain why horticultural sand, used for container-grown plants, need to be 'acid washed'.

Exercise 15.4

Sulphur requirements to lower pH

Background

Once the free lime has been neutralized and for non-calcareous soils which have less than 0.1 per cent free lime, the quantities of sulphur required to lower the pH are dependent upon the soil texture and starting pH. Table 15.2 illustrates this relationship.

Refer to the soil soil texture exercises for the texture types that the labels represent in the table.

Table 15.2 Sulphur requirement to lower pH by one pH unit (t/ha)

Soil texture	S	LS	SL	SZL	ZL	SCL	CL	ZCL	SC	ZC	C
Sulphur requirement	0.75	1.0	1.0	1.5	1.8	2.2	2.2	2.2	2.7	2.7	2.7

Aim

To assess the quantity of sulphur required to lower soils of varying texture and pH.

Method

Using Table 15.2 answer the following questions.

Results

Enter your results in the table provided.

State the amount of ground sulphur required to lower the pH of:	Sulphur requirement (ground sulphur)	
	t/ha	g/m²
1. A seaside, sandy silt loam, bowling green from pH 7.0 to pH 6.0		
2. A waterlogged clay bedding plant area from pH 7.5 to pH 6.5		
3. A loamy sand planned rhododendron unit from pH 7.0 to pH 5.0		
4. An ericaceous unit (silt loam) from pH 6.5 to pH 5.0		
5. Allotments (silty clay loam) from pH 7.0 to pH 6.5		
6. An unused calicole species demonstration site (sandy loam) from pH 8.5 to pH 6.5		

Conclusions

State how much sulphur is required to lower the pH of Sample Soil A and B of Exercise 15.3 to pH 5.5.

Exercise 15.5

Lowering pH – integration of skills

Background

A poorly drained local common is being converted into a rhododendron and ericaceous heather park. Sample soils A and B, have been taken from this common for analysis. Calculate, using the skills developed during these exercises a sulphur requirement to lower the pH to pH 5.5.

Aim

To integrate and strengthen the skills developed in these exercises to calculate a sulphur requirement for a range of situations.

Method

Investigate soils A and B to complete the table provided.

Results

Enter your results in the table provided.

Growth medium	Soil texture	Free lime content $CaCO_3$ (%)	Sulphur required to neutralize free lime $CaCO_3$ (t/ha)	Present pH	Desired pH	Additional ground sulphur to lower pH (t/ha)	Total ground sulphur required (t/ha)
Example soil	clay loam	20%	20	7.5	5.5	$2.2 \times 2 = 4.4$	$20 + 4.4 = 24.4$
Sample soil A							
Sample soil B							

Conclusions

1. Why is it unwise to apply sulphur to areas where there are established plants growing?
2. How should sulphur be applied?
3. What other materials can be used to increase soil acidity in place of sulphur?

Answers

Exercise 15.1. Soil and compost pH testing

Results

Soils with a pH above pH 7.0 should be tested.

Compost A = pH 4.5–5.0.

Compost B = pH 5.5.

Compost C = pH 7.0–7.5.

Conclusions

1.   The soil with the highest pH.

2.   Choice from: rising water tables, high dissolved nutrient content in irrigation water, poor drainage, or poor watering practice, such as excessive drying out between irrigations.

3.   An acid-based compost should be used and mixed into the soil surrounding the rootball prior to planting.

Exercise 15.2. Soil texture

Results

A sandy soil texture should be used for soil A and a clayey texture for soil B.

Conclusions

1.   Soil B because it has more clay.

2.   Soil B because it has a greater buffering capacity.

3.   Suitable plants should be selected from Table 13.2.

Exercise 15.3. Free lime (calcium carbonate) content

Results

Typical results are:

sea shells 10 per cent $CaCO_3$

beach sand 10 per cent $CaCO_3$

grit 10 per cent $CaCO_3$

silver sand 0.1–0.5 per cent $CaCO_3$

horticultural sand 0.1 per cent $CaCO_3$

vinegar never gives better than 0.1 per cent regardless of the true value

Conclusions

1.   Vinegar is not a suitable substitute for 10 per cent hydrochloric acid yet it has developed something like mythical status among practitioners in the industry.

2.   Sea shells are mostly composed of calcium carbonate and so would only raise pH further. However, they might instead be used as a liming material.

3.   The sample giving the strongest reaction.

4.   Horticultural sand is acid washed, usually in low concentrations of citric acid, to neutralize the free lime. In this way the sand is then suitable to be mixed with compost without affecting the pH of the media.

Exercise 15.4. Sulphur required to lower pH

Results

1.   1.5 t/ha, 1500 g/m^2.

2.   2.7 t/ha, 2700 g/m^2.

3.   2 t/ha, 200 g/m^2.

4.   2.7 t/ha, 270 g/m^2.

5.   1.1 t/ha, 110 g/m^2.

6.   2 t/ha, 200 g/m^2.

Conclusions

The question requires evidence that the candidates can interpret Table 15.2 correctly for each sample soil.

Exercise 15.5. Lowering pH – integration of skills

Results

Soils A and B should first be textured, then the pH determined. Next use Table 15.1 to determine the free lime and sulphur required, then use Table 15.2 to determine the additional sulphur required to lower pH to the target of pH 5.5.

Conclusions

1.   The sulphur changes to sulphuric acid which might destroy plant roots.

2.   Several months before planting.

3.   Choice from: ferrous sulphate, aluminium sulphate, low-grade pyrites and calcium sulphate (gypsum) and commercial products such as Phostrogen.

Chapter 16   Soil organisms and composting

Key facts

1. Soil organisms are essential to recycling nutrients.
2. Soil organisms vary in size from microscopic to earthworms.
3. Decomposers are scavengers that blend into small pieces and digest organic matter.
4. There are many ways to classify soil organisms, e.g. scavengers and decomposers, or by size: micro, meso and macrofauna.
5. The rhizosphere is the name for the layers in the soil in which plant roots and soil organisms interact.

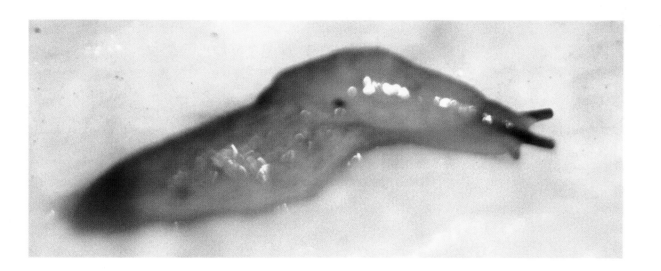

Background

The soil is teaming with life, although because many organisms cannot be seen with the naked eye, most people are never aware that any life exists. In $1\,m^2$ of woodland soil, there are estimated to be:

- 1000 species of animals
- 10 million nematode worms and protozoa
- 500 000 mites and springtails
- 10 000 other Invertebrates.

In $1\,cm^3$ of soil there are estimated to be:

- 6–10 million bacteria
- 1–2 km of fungal filaments (hyphae).

Functions of soil organisms in nutrient recycling

Survival of soil organisms is influenced by the interactions between them. Dead, decaying plants and their residue feed these organisms. They have a key role to play in healthy plant growth and control of soil-borne diseases. They decompose organic matter, recycling nutrients and improving soil structure. Soil organisms are important in horticulture as they are active competitors in decomposing organic matter to produce nutrient-rich humus. As soon as a leaf falls to the ground it is subjected to a coordinated attack by soil organisms. Bacteria and fungi start this attack and are followed by mites, snails, beatles, millipedes, woodlice, and earthworms. (Soil organisms are also pests consuming growing plants and increasingly are used as agents of biological control, preying on one another.) These organisms are referred to as 'decomposers' because their job is to decompose dead organic matter, and recycle the nutrients back into the soil for other plants to use.

The role of decomposers is to:

- recycle nutrients
- increase aeration
- stimulate microbial decomposition.

Without decomposers raw materials (nutrients) would run out. Decomposers have the greatest biomass in the system equivalent to the weight of about forty horses per hectare.

Exercise 16.1

Classification of soil organisms

Background

The term **flora** refers to vegetative material.

Fauna refers to animal organisms. The largest animals (macrofauna) feed on leaf/compost litter. The mesfauna (e.g. mites) help circulate nutrients between the litter layer and humus formation. Acting together these organisms mash, digest and oxidize all forms of organic matter including fallen leaves, trunks, dead grass, faeces, defunct bodies and some species even devour one another.

The soil zone in which plant roots and soil organisms interact is called the rhizosphere. Soil organisms may be grouped into categories according to their size and their efficiency at chopping up and decomposing organic matter. This helps to reveal the functional role of decomposers in soil.

Microflora

- bacteria
- fungi.

Microfauna

These are predators of fungal hyphae and bacteria:

- protozoa (amoeba, flagellates, ciliates)
- nematodes (millions per square metre).

Mesofauna

These attack plant litter and the recycled faeces of other soil animals:

- springtails
- mites
- enchytaeid worms
- some fly larvae
- small beetles.

Macrofauna

These are litter feeding invertebrates. They act as mechanical blenders breaking up organic matter and exposing fresh organic surfaces to microbes:

- millipedes
- centipedes
- woodlice
- spiders
- earthworms
- beetles
- slugs
- snails
- harvestmen.

Aim

To gain familiarity with both the range and classification of soil organisms involved in decomposition and nutrient recycling.

Apparatus

Figure 16.1

Useful websites

http://soils.usda.gov

www.soil-net.com

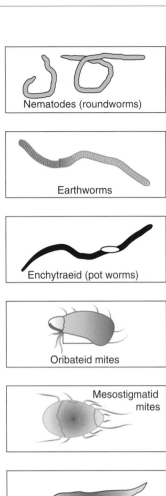

Nematodes (roundworms)

Bacteria, fungi, algae, living plant roots (eelworms). Some are predators of soil animals

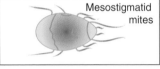

Earthworms

Bacteria, fungi, decaying plant matter

Enchytraeid (pot worms)

Fungi and nematodes

Oribateid mites

Bacteria, fungi, decomposing plant remains

Mesostigmatid mites

Earthworms, potworms, nematodes and mites. Many are parasites of insects used in biocontrol

Turbellaria (flatworms)

Microscopic animals in the water film

Pseudo-scorpions

Predators of mites and springtails

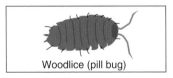

Spiders

Predators of most insects and small organisms

Woodlice (pill bug)

Feed on dead plant and animal matter

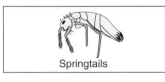

Springtails

Bacteria, fungi and decaying organic matter

Beetles

Larvae can be very damaging (e.g. wireworms). Consume green leaves of growing pot plants

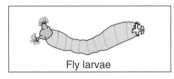

Fly larvae

Feed on decaying or living plants, underground roots and stems (e.g. leatherjackets)

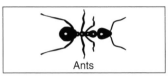

Ants

Consume juvenile soil animals and some plant remains

Moths and butterflies

Mostly non-feeding pupae, but some larvae feed and live in the soil

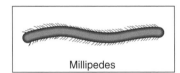

Millipedes

Consume decaying plants and living roots. Important in the mechanical breakdown of humus

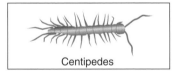

Centipedes

Consume juvenile soil animals

Slugs and snails

Soil surface feeders of fungi and living plants. Some are carnivorous on earthworms

Grasshoppers

Roots, stems and small insect feeders

Figure 16.1 Soil organization and feeding profile

Method

1. For each organism listed in Figure 16.1 state whether it is micro-, meso- or macrofauna.
2. Give an example of:
   (a) its role in decomposition
   (b) an organism which may decompose it.

Results

State your results in the table provided.

Soil organism classification

Micro, meso or macro fauna	Organism	Type of organism to decompose	Type of organism decomposing it
	Snails		
	Mites		
	Springtails		
	Beetles		
	Pseudo-scorpion		
	Millipedes		
	Ants		
	Earthworms		
	Centipedes		
	Spiders		
	Fungal mycellum		
	Slugs		
	Woodlice		
	Larvae		
	Nematodes		
	Bacteria		
	Roundworms		

Conclusions

1. Which classification grouping appears to have the largest population size?
2. Which grouping has the most diverse range of organisms?
3. Distinguish between the terms 'fauna' and 'flora'.

Exercise 16.2

Biodiversity in a compost heap

Background

Biodiversity refers to the variety of soil and vegetative organism species present.

In a rotting compost heap there will be a tremendous range of organisms present as they are specialized to particular jobs which ensure that all the material is digested and recycled in the form of humus. The macrofauna, for example, will be cutting and blending larger pieces of vegetation which in turn will expose fresh tissue for bacterial and fungal attack. The range of organisms present may be collected in a jar called a pooter (see See Chapter 21). This allows you to suck organisms into the jar, but avoids them entering your mouth. Specimens can then be studied either on site or in the laboratory later.

Aim

To investigate and quantify the diversity of decomposing organisms in a compost heap and observe changes with depth.

Apparatus

Figure 16.1
Pooters
Spades
Trowel
Sample bags
Meter rules

Method

Cover all wounds to reduce risk from infection.

1. Select a large, well-rotted compost heap or similar area with decomposing organic matter (e.g. a woodland floor).
2. Visually isolate 1 m^2 of the heap for investigation.
3. Using the pooters collect and count soil organisms on the surface of the heap.
4. Slowly excavate the heap to depths of 15 cm, 30 cm 50 cm and 1 m, recording organisms present.

Results

Enter your results in the table provided.

Compost heap organism biodiversity results						
Depth	Organisms present					
	Microfauna		Mesofauna		Macrofauna	
	Type	No.	Type	No.	Type	No.
Surface						
15 cm						
30 cm						
50 cm						
1 m						

Conclusion

1. Which group of organisms had the largest population size?
2. Express each classification group as a percentage of the total.
3. Explain how the diversity of organisms changed with depth.
4. What percentage of the total organism population, was each organism:
   (a) at each depth?
   (b) per cubic metre?
5. Can your results be explained in terms of the type of organic matter present? If so, how?

Answers

Exercise 16.1. Classification of soil organisms
Results

Classification	Organism	Decomposes	Decomposed by
macro	snails	fungi loving plants	birds, centipedes, hedgehogs
meso	mites	nematodes, worms, algae, lichen	ants, beetles
meso	springtails	algae, bacteria, fungi	pseudo-scorpions
macro	beetles	spiders, mites, beetles	larger carnivores
meso	pseudo-scorpions	mites, springtails	ants
macro	millipedes	decaying plants	moles, birds, hedgehogs
macro	ants	spiders, centipedes, mites	hedgehogs, birds
micro	fungal mycelium	dead/decaying plants and animals	nematodes, slugs, snails
macro	slugs	fungi and living plants	birds
macro	woodlice	dead plants and animals	hedgehogs
meso	larvae	living and decaying plants	birds, hedgehogs
micro	nematodes	bacteria, fungi	mites
micro	bacteria	leaf and animal cells	some fungi, protozoa, springtails
micro	roundworms	bacteria, fungi	mites

Conclusions

1. Micro, because they are nearer the bottom of the food chain.

2. Macrofauna.

3. Fauna refers to animals, flora to vegetation.

Exercise 16.2. Biodiversity in a compost heap
Results

Results will vary with the type of vegetation and location of the site investigated. A range of organisms as outlined in Figure 16.1 should be observed.

Conclusions

1. Microfauna

2. Percentage should be calculated using the formulae: number present/total number recorded × 100

3. With increasing depth diversity falls and microfauna dominate.

4. Percentages should be calculated using the formulae in 2.

5. There may be simple correlations between vegetation type and organisms. For example, beetles prefer wood and leaf litter, fungus readily colonizes grass clippings.

Chapter 17   Soil organic matter

Background

Organic matter in the soil is not a well-defined property. It consists of a wide range of compounds forming a biochemical continuum that ranges from single-celled organisms to higher plants and microbes to material of animal origin. Thus it is plant and animal material that is decomposing. Fully decomposed organic matter is called humus.

In simple terms organic matter may be considered as all the organic material in soil which passes through a 2 mm sieve.

Examples of the use of organic matter in horticulture include:

- peat
- farmyard manure (FYM)
- straw incorporation
- mulching.

Organic matter (OM) is substances derived from the decay of leaves and other vegetation, including animal waste and tissues. It may have been broken down into extremely fine humus. Adding organic matter to soil adds both oxygen from turning the soil and recycled carbon – which stimulates the growth of beneficial bacteria and micro-organisms – together with other nutrients that aid healthy plant growth. The organic matter content is now recognized as a key indicator of sustainability and soil quality.

Exercise 17.1

Soil organic matter determination

Background

Organic matter can be approximately determined by measuring the organic carbon content of sieved soil. Organic matter itself ranges from 40 to 60 per cent carbon. When the soil is heated the organic matter burns away (loss on ignition). The mineral fraction is unaffected by the heat; the change in weight therefore approximates to the organic matter content.

Aim

To determine the organic matter status of different soils.

Apparatus

Analytical balance	Weigh boats
Metal crucible	Heat-resistant mat
Bunsen burner	Tripod
Forceps	Spatula
Calculators	Goggles
Dry sieved sample A (lawn soil)	Dry sieved sample B (orchard herbicide strip soil)

Useful website

www.sassa.org.uk

Method

You may find it helpful to draw a labelled diagram of the apparatus.

1. Collect soil A and soil B from a lawn and an orchard herbicide strip respectively.
2. They must be air-dried, ground and sieved to pass a 2 mm sieve.
3. Zero the balance and weigh the weigh boat (a).
4. Weigh out exactly 50 g of soil in a weigh boat (b).
5. Place soil in metal crucible.
6. Prepare your furnace by positioning a tripod over a heat-resistant mat and locating a Bunsen burner beneath.
7. Using two Bunsen burners heat the tray from below as strongly as possible for 15 minutes. The tray and soil may become red hot.
8. Allow the tray to cool. Tip the soil into the weigh boat and record the new weight (d).
9. Calculate OM content using the formula given in the results table (g).

Results

Enter your results in the table provided. The loss in weight represents the organic matter that has been burned off. It should be expressed as a percentage of the weight after heating.

Determination	Soil A	Soil B
(a) weight of weigh boat		
(b) weight of weigh boat and fresh soil		
(c) Fresh weight of soil		
(d) weight of weigh boat and soil after heating		
(e) weight of soil after heating		
(f) change in weight of soil		
(g) % organic matter content $\dfrac{\text{Change in weight (f)}}{\text{Weight of soil after heating (e)}} \times 100$		

Conclusions

1. Explain why is the soil sample should be sieved.
2. What sieve size is used?
3. If the soil sample had not been completely dry to start with, how would this have affected your results?
4. What is the reason for heating the soil so strongly and why should this result in a loss in weight?
5. It is possible that 15 minutes heating is not sufficient to achieve the maximum loss in weight of the soil. What would you have to do to be quiet sure that the maximum weight loss had been achieved?
6. Which soil had the greatest organic matter status?

7. Describe how the variations in organic matter status reflects the land use in these two situations.

8. State two benefits from organic matter incorporation into a soil.

Exercise 17.2

Properties of organic matter

Background

Organic matter is one of the most important ingredients in a soil and is a key indicator of quality and sustainability. Some of the benefits of organic matter in the soil include:

- high cation exchange capacity (ability for chemical transfers)
- increased water-holding capacity (e.g. 80–90 per cent of weight)
- improved structure and stability for roots growth
- dark colour, which absorbs sunlight, heating up quickly in the spring and speeding up germination
- improved response to fertilizers
- improved resistance to erosion
- increased soil nutrients through recycling
- buffer the release of pollution into air and water.

For these reasons horticulturalists should try to maintain a high level of organic matter in the soil. Organic matter can increase the water-holding capacity of soils. Therefore if organic matter is added to soil, less water will drain away because the volume of water held at field capacity will be artificially higher.

Aim

To demonstrate the increased water holding capacity of soils with organic matter incorporation.

Apparatus

Capillary Tubes A and B
Sandy soils
Retort stand
Beaker
Measuring cylinder

Useful websites

www.defra.gov.uk

Method

You may find it useful to draw a diagram of the apparatus.

1. Mix some organic matter (e.g. peat, farmyard manure or straw) with soil A.
2. Fill one capillary tube with soil A the other with soil B, and fix into retort stands with a beaker beneath each.
3. Pour 100 ml of water into the top of each tube and collect the drainage water in a measuring cylinder.
4. Record the volume of drainage water collected from soils A and B.

Results

Enter your results in the table provided in the next page.

Sample	Volume of water added	Volume of drainage water collected	Interpretation
Soil A			
Soil B			

Conclusion

1. Explain which soil has the highest organic matter content by reference to your results.
2. Why is organic matter essential in a fertile soil?
3. Name three types of bulky organic matter commonly added to soil.
4. State two other ways in which organic matter can improve the soil.

Exercise 17.3

Organic matter and nitrogen

Background

The breakdown of organic matter liberates nutrients for use by other plants (nutrient cycles).

Nitrogen is one such nutrient supplied in this way. Nitrogen is the most important nutrient in plant growth, but is suspected of causing cancer (carcinogen), and is appearing in drinking water in excess of the EEC limit of 50 mg/l.

Nitrogen cycle

Figure 17.1 shows the stages of the nitrogen cycle from atmosphere to soil solution and back again.

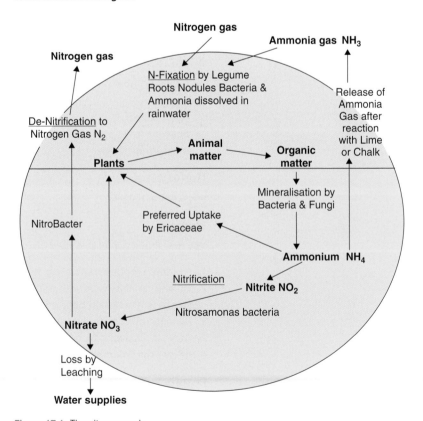

Figure 17.1 The nitrogen cycle

Nitrogen is a gas making up approximately 79 per cent of the atmosphere, but plants cannot use gaseous nitrogen. Most plants obtain nitrogen from the soil solution in the form of nitrates and to a less extent ammonium. Plants use nitrogen to form proteins such as chlorophyll.

The nitrogen in organic matter occurs in complex forms which are insoluble in water and cannot be directly used by plants. To make it available it must be converted to nitrates. This conversion (nitrification) is brought about by bacteria in three distinct phases. First, decomposition of organic matter results in the production of ammonium gas (ammonification) brought about by fungi and ammonifying bacteria. Second, ammonium converted by nitrosomonas-nitrifying bacteria, into nitrite. This form of nitrogen is toxic to plants. Third, nitrobacter converts nitrite to nitrates, which are the main form of nitrogen for plant growth This requires aerobic soil conditions. In anaerobic conditions, such as on waterlogged land, denitrification occurs. This is the conversion of nitrates into nitrogen gas.

Nitrogen fixation

Although plants cannot obtain their nitrogen directly from the air, some bacteria can. For example, rhizobium bacteria live in root nodules on some legumes and trap atmospheric nitrogen. They live in a symbiotic relationship with plants exchanging nitrates for sugar solution made during photosynthesis.

Other organic matter organisms, including azobacter and clostridia, are also able to trap atmospheric nitrogen. When they die and decompose, their proteins can be recycled into nitrates for other plants to use.

Humans greatly impact the nitrogen cycle through the manufacture of fertilizers and burning fossil fuels. The nitrogen in the seas and oceans has increased due to nitrate runoff and sewage disposal. This can cause eutrophication, leading to algal blooms in ponds and streams. The resulting loss of oxygen damages fish and other ecosystem life.

In this exercise two soils of different organic matter content are compared. The glasshouse soil will have been subjected to intensive cultivation and will be very high in organic matter content from dead roots and vegetative matter which should be clearly visible. The grassland or herbicide strip soil will have less organic matter, particularly if a sample is taken from the subsoil. Nitrogen exists in several different forms in the soil including nitrate, nitrite and ammonium. Nitrate is the most soluble form and the form in which most plants take up nitrogen through their roots. In this exercise all the nitrate that leaches is assumed to have come from the breakdown of organic matter rather than from fertilizer residues.

Aim

To investigate the effect of organic matter on nitrate availability and leaching.

Apparatus

Funnels	Clamps
Beakers	Measuring cylinder
Distilled water	Retort stand
Cotton wool	Soil A (glasshouse soil)
Soil B (grassland/herbicide strip soil)	Merkoquant nitrate test strips

Method

You may find it helpful to draw a labelled diagram of the apparatus.

1. Place a small cotton wool ball in the neck of a funnel, held in the retort stand.
2. Repeat the above a second time.
3. Add Soil A and Soil B to different funnels.
4. Place a beaker beneath the funnel.
5. To soils A and B, slowly add 100 ml additions of distilled water until the first drops of leachate (drainage water) appear.
6. Place a nitrate test strip in the leachate for 1 second, then after a further 60 seconds, (after which time full colour development will be complete) compare the colours against the standards colour chart provided with the test strips to ascertain the nitrate content.
7. Pour away the remaining leachate and rinse the beaker with distilled water before replacing.
8. Continue to add further 25 ml additions, until an additional 200 ml has been added, discarding the leachate between additions.
9. Measure the nitrate content as in step 6.

Results

Record your results in the table provided.

Soil type	Nitrate content (mg/l)		% change
	leachate 1	leachate 2	
Soil A (glasshouse)			
Soil B (grassland/herbicide strip)			

Conclusions

1. Which soil had the highest nitrate content?
2. How can your results be explained by reference to the organic matter content of each soil?
3. Which soil is more susceptible to nitrate leaching?
4. Comment on how organic matter content may be affected by earthworms and cultivations.

Answers

Exercise 17.1. Soil organic matter determination

Results

Great variation in results has been observed, typical values are 15–20% OM for lawns and 7–10% for bare soil/herbicide strips.

Conclusions

1. OM is defined as that which passes through a 2 mm sieve. Larger pieces of OM such as plant roots etc., are not considered in this determination.

2. 2 mm sieve.

3. The results would be excessively high because they would also include the loss of weight from the vaporization of water.

4. Strong heat turns the OM into carbon gas which escapes into the atmosphere. The loss is recorded as a reduction in weight.

5. Laboratory methods require temperatures of over 300°C to be achieved in a purpose-built oven. Bunsen burners cannot achieve this temperature although heating for longer than 15 minutes increases the accuracy of results.

6. The lawn, due to the massive amounts of fibrous roots and root hairs produced in the topsoil.

7. Turf always produces a mass of fibrous roots in the topsoil. These are constantly growing and decaying producing high OM status. Bare soil/herbecide strip have no plants growing in them, or might typically have orchard trees with bare sol beneath and so produce lower OM levels.

8. Choice of two from: high cation exchange capacity, increased water holding capacity, improved structure and stability, dark colour, which absorbs sunlight, heating up quickly in the spring, and improved response to fertilizers.

Exercise 17.2. Properties of organic matter

Results

Soil A retains far more water than soil B.

Conclusions

1. Soil A had the greatest OM content. This is indicated in the fact that less water drains through the capillary tube. A characteristic of OM is that it can increase the water-holding capacity of a soil.

2. Organic matter is essential to soil fertility because it breaks down into nutrients essential for plant growth. Nitrogen in particular is recycled from OM.

3. Choice from: peat, farmyard manure, straw, and mulching.

4. Choice from: high cation exchange capacity, increased water-holding capacity, improved structure and stability, dark colour, which absorbs sunlight, heating up quickly in the spring, and improved response to fertilizers.

Exercise 17.3. Organic matter and nitrogen

Results

The glasshouse soil will contain more nitrate than grassland soil.

Conclusions

1. Soil A. The glasshouse soil benefits from fertilizer additions yet does not have winter rainfall to leach away nitrates.

2. Glasshouse soils are intensively cropped, e.g. all year round chrysanthemums, producing masses of roots and root hairs constantly renewing and decaying to release nitrate. By comparison grassland has a seasonal growth pattern and is dormant for part of the year.

3. Soil B since it will be subjected to periodical heavy rainfall.

4. Earthworms create aeration channels and bury OM, leading to decay. Cultivation practices such as rotavation turn the soil and stimulate microbial decomposition of OM.

Chapter 18   Plant nutrition

Key facts

1. A nutrient is an element essential for plant life.
2. There are three types of nutrients: major, minor and trace elements.
3. Nutrients are required in balanced proportions.
4. An over-application of trace elements is toxic to plants.
5. Nitrogen is required in large quantities but easily leaches away.
6. Nitrogen levels are estimated based on summer rainfall, soil texture and previous plants grown.

Background

As with all living things, plants need food (nutrition) for their growth and development. Plants live, grow and reproduce by taking up water and mineral substances from the soil, together with carbon dioxide from the atmosphere, and energy from the sun, to form plant tissue.

A large number of elements are found in plant tissue. Of these, sixteen have been found to be indispensable – without these, the plant would die. These elements are referred to as nutrients.

Nutrients from the atmosphere or soil water

Carbon (C)
Hydrogen (H)
Oxygen (O)

Nutrients from the soil/growth medium

Nitrogen (N)
Phosphorus (P)
Potassium (K)
Calcium (Ca)
Magnesium (Mg)
Sulphur (S)
Iron (Fe)
Zinc (Zn)
Manganese (Mn)
Copper (Cu)
Boron (B)
Molybdenum (Mo)
Chlorine (Cl)

Plants have also been found to benefit from supplies of cobalt (Co), sodium (Na), and silicon (Si), but these are not considered 'essential' nutrients. That is, without them, the plant will not die.

The atmospheric gases (C, H, and O) are essential for photosynthesis. They are also used in making plant carbohydrates and proteins.

Of the soil-derived nutrients nitrogen, phosphorus and potassium are used in the largest quantities followed by magnesium, calcium and sulphur. This may be as little as 1 kg/ha up to several 100 kg/ha, and they are therefore referred to as **major nutrients** (or macronutrients).

The remaining nutrients (Fe, Cu, Bn, Mo, Cl, Mn, and Zn) are required in smaller quantities and in large amounts are toxic to plants. These are referred to as **trace elements** (or micronutrients).

The soil or growth media must contain all the essential nutrients in sufficient quantity and in balanced proportions. These nutrients must also be present in an **available** form before plants can use them. An **inadequacy** of any one of these elements will inhibit plants from growing to their full potential.

Exercise 18.1

Plant nutrition

Aim

To develop knowledge of the ingredients of plant nutrition.

Apparatus

Background text section

Method

Read the following questions and phrase suitable answers.

Results

1. What is a plant nutrient?
2. How many plant nutrients are there?
3. Explain what a major nutrient is.
4. List the major elements.
5. Explain what a trace element is.
6. List the trace elements.
7. What effect will an over-application of trace elements have on plant growth?
8. What other terms are used to describe:
   (a) major elements?
   (b) trace elements?
9. Which nutrients are mostly supplied as gases?

Conclusions

1. What nutrition is required to support plant growth?
2. Explain what effect an absence of a plant nutrient will have on plants?
3. Explain the impact of an inadequate supply of any nutrient on growth?
4. Why are sodium, silicon and cobalt not 'technically' plant nutrients?

Exercise 18.2

Tomato plant deficiency symptoms

Background

The ability to correctly identify nutrient disorders is extremely helpful in diagnosing plant health problems.

The problems will mostly arise in pot plants where the compost may have been mixed on the nursery or where supplementary overhead irrigation also supplies some nutrition. Typically, a fertilizer such as phosphorus will have been omitted from the compost mix and the plants will grow with purple leaf edges or other symptom of the deficiency. Figure 18.1 shows an example of phosphorus deficiency in a pot plant. In soils, over-cropping

Figure 18.1 Symptoms of phosphorus deficiency

and soil texture variations may induce a deficiency. In either case, the problem must be identified and corrective action taken.

In this exercise tomato seeds are sown in an inert material (such as rockwool or perlite) and watered using a solution which lacks one nutrient. Tomatoes are a particularly good plant to grow since they demonstrate very easily the deficiency symptom under study.

The plant grown with a complete range of nutrients should demonstrate healthy growth and colouring. Tap water contains some nutrients but in very low quantities. Distilled water is the purest form of water available and will contain no nutrients at all. The tomatoes grown in this will therefore have to survive just on the energy contained in the germinating seed. The remaining plants will show varying degrees of leaf colour, growth, root development and stem strength which should all be observed.

Aim

To gain increased competence at the visual diagnosis of nutrient deficiency symptoms.

Apparatus

> Tomato seeds
> Rockwool cubes or perlite growing medium

Tomato plants liquid fed solutions lacking: nitrogen, phosphorus, potassium, magnesium, calcium, sulphur, sodium and iron (e.g. The 'Long Ashton Water Culture Kit', from Griffen and George Suppliers, provides premixed chemicals that are dissolved in water).

Tomato plants liquid fed with tap water and distilled water.

Tomato plant fed with complete range of nutrients.

Useful websites

www.primalseeds.org/nutrients.htm

www.biotopics.co.uk/plants/plantm.html

Method

You may find it useful to draw a diagram of the deficiency symptoms shown in the plants.

1. Sow half a dozen seeds in rockwool cubes (or perlite) on the same day (enough for each of the above treatments).
2. Water each seed and subsequent germinating plant with one of the above solutions.
3. If the cube becomes crowded, thin out the plants, leaving those that demonstrate good symptoms of the deficiency under study.
4. After several weeks, compare the growth profiles of the tomato plant receiving the complete feed with those lacking N, P, K, Mg, and those fed with tap water and distilled water, recording the visual symptoms of the deficiency.

Results

Record your results in the table provided in the next page.

Type of feed	Description of condition and symptoms of deficiency
Complete nutrient range	
Distilled water	
Tap water	
Lacking N	
Lacking P	
Lacking K	
Lacking Mg	

Conclusion

1. Why was distilled water used in this experiment?
2. Explain why you would not rely on tap water as a nutrient supply, especially when feeding cut flowers in a vase.
3. Which nutrient deficiency is most easily recognized?
4. Assuming that the deficiency symptoms that you observed are representative of the typical symptoms likely to be found on a range of plants, identify the deficiencies present in each of the following situations:
   (a) A *Chrysanthemum* suffering from stunted growth and light green leaves.
   (b) *Poinsettia* containing yellow leaf edges and pale green/yellow blotches between leaf veins. The leaves are small and brittle, turning upwards at their edges, and suffer premature leaf drop.
   (c) *Alstroemeria* having a scorched yellow look to the edges of the older leaves, stalks are weak and the plant collapse (go floppy) easily.
   (d) *Solanums* with leaf edges containing a thin margin of purple colouring.

Exercise 18.3

Function of plant nutrient

Aim

To recognize the physiological importance of nutrients to plant development.

Apparatus

List of nutrients C, H, O, N, P, K, Mg, Ca, S, trace elements.
Tomato plant specimens from Exercise 18.2
Table 18.1

Table 18.1 The role of plant nutrients and their deficiency symptoms

Nutrient	Role in plant	Susceptible soils	Deficiency symptoms
Carbon and hydrogen	• Component of all carbohydrates and proteins • Essential in photosynthesis	Supplied as atmospheric gases and in the soil solution	Deficiencies unknown
Oxygen	• Component of carbohydrates and proteins • Essential in photosynthesis and respiration		
Nitrogen[a]	• Essential for vegetative growth • Protein and chlorophyll formation	Light-textured soils lacking organic matter	• Stunted growth • Light green to pale yellow leaves starting at the leaf tip, followed by death of the older leaves if the deficiency is great • In severe cases the leaves turn yellowish-red along their veins and die off quickly • Flowering is both delayed and reduced
Phosphorus[a]	• Root growth during the early stages of plant life • Helps plant to flower and fruit productively • Enables meristematic growth • Component of amino acids and chlorophyll • Necessary for cell division	Occasionally found on heavy-textured soils and peat composts	• Bronze to red/purple leaf colour • Shoots are short and thin • Older leaves die off rapidly • Flower and seed production are inhibited • Restricted root development • Delayed maturity
Potassium[a]	Essential for many physiological reactions within the cell: • Control of osmosis • Resistance to drought, disease and frost • Influences the uptake of other nutrients e.g. Mg • Flower and fruit formation • Improves the quality of seed, fruit and vegetables	Occasionally found on light-textured soils	• Stunted growth • Scorched look to the edges of older leaves (chlorosis) gradually progressing inwards • Stalks are weak and plants collapse easily • Shrivelled seeds or fruit • Brown spots sometimes develop on leaves

(Continued)

Table 18.1 (Continued)

Nutrient	Role in plant	Susceptible soils	Deficiency symptoms
Magnesium[b]	• Essential for chlorophyll formation • Influences phosphorus mobility • Influences potassium uptake by the roots • Helps the movement of sugars within the plant	Often found on light-textured, sandy or peaty soils, especially in high rainfall areas where there is an excess of calcium or potassium	• Leaves develop yellow margins and pale green/yellow blotches between veins (interveinal chlorosis) • Blotches become yellower and eventually turn brown • In final stages leaves are small and brittle, edges turn upwards • In vegetables, plants are coloured with a marbling of yellow with tints of orange, red and purple • Stems are weak and prone to fungus attack • Premature leaf drop
Calcium[b]	• Essential for development of growth tissue (tips) • Constituent of cell walls • Essential for cell divisions, especially in roots • Maintenance of chromosome structure • Acts as a detoxifying agent by neutralizing organic acids in plants • Firmer fruit • Prevents bitter pit in apples and cork spot in pears • Improves storage life	Normally, if the pH of the soil is satisfactory, then so is calcium content. Deficiencies are usually rare but can be found on light-textured, acid or peaty soils	• Deficiencies are not often seen, partly because secondary effects associated with high acidity limit growth • Chlorosis of the young foliage, and white colouration to edges • Death of the terminal bud • Growing point may shrivel up and die • Grey mould (botrytis) infection to growing point • Distorted leaves with the tip hooked back • Death of root tips and damaged root system appearing rotted • Buds and blossoms shed prematurely • Stem structure weakened
Sulphur[b]	• Component of amino acids, proteins and oils • Involved with activities of some vitamins • Aids the stabilization of protein structure	Sulphur is a common atmospheric pollutant and therefore rarely added as fertilizer	• Uniform yellowing of new and young foliage
Trace elements	• Many functions mainly associated with photosynthesis, nitrogen assimilation or protein formation	Rare in soils but more common in composts	• Various symptoms including brittle leaves, chlorosis, leaf curling poor fruiting, and distorted leaf colours

[a] Nitrogen, phosphorus and potassium are the three most important macro-elements and are referred to as the 'primary major nutrients'

[b] Magnesium, calcium and sulphur are the secondary major nutrients, but are not required in the same quantities as N, P or K

Method

Observing the tomato plants and the literature match the following list of nutrient functions with the nutrient responsible.

Results

Which nutrient:

(a) increases the vegetative growth?

(b) is a component of carbohydrate, proteins and essential in photosynthesis (two nutrients)?

(c) is a component of carbohydrate and proteins and is essential in photosynthesis and respiration?

(d) controls osmosis and resistance to disease?

(e) is a constituent of chlorophyll?

(f) improves fruit storage life and constituent of cell walls?

(g) is a component of some amino acids?

(h) has many functions mainly connected with photosynthesis, nitrogen assimilation and protein formation?

(i) is responsible for root development in young plants?

Conclusions

1. Explain the difference between a major and minor nutrient.

2. How are primary major nutrients distinguished from secondary major nutrients?

3. Which nutrients most commonly need to be applied as fertilizers?

Exercise 18.4

Major nutrient roles

Background

The primary major nutrients are the nutrients that most often suffer deficiency and will need to be applied as fertilizers. It can be useful to remember their role in the plant by reference to a graphical representation of a plant with the function of the nutrient closely associated with its position on the plant diagram.

Aim

To consolidate learning of the role played by nutrients in plant physiology.

Apparatus

Figure 18.2
Table 18.1

Useful website

www.fao.org/ag/AGP/AGPC/doc/publicat/FAOBUL4/FAOBUL4/B402.htm

Method

Observe the plant illustrated in Figure 18.2 and decide which nutrients are being described in the text.

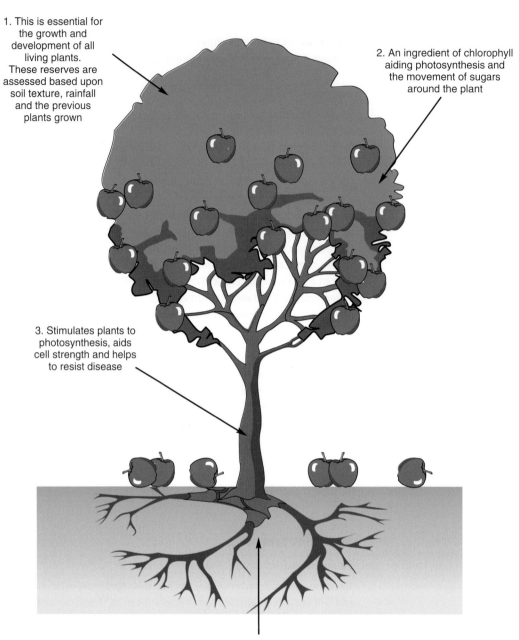

1. This is essential for the growth and development of all living plants. These reserves are assessed based upon soil texture, rainfall and the previous plants grown

2. An ingredient of chlorophyll aiding photosynthesis and the movement of sugars around the plant

3. Stimulates plants to photosynthesis, aids cell strength and helps to resist disease

4. This encourages root development and enables plants to flower and fruit productively

Figure 18.2 Plant nutrient roles

Results

Enter your answers below.

1. ....

2. ....

3. ....

4. ....

Conclusion

Define the terms 'major element' and 'minor element'.

Nitrate fertilizers

Background

> Nitrogen is probably the most important nutrient.

The deficiency is fairly common, especially on light-textured soils lacking organic matter. It is used in protein and chlorophyll formation and is very effective at stimulating vegetative growth. Most plants prefer to take up nitrogen in the form of nitrate (NO_3), but Ericaceae species prefer it as ammonium (NH_4). Nitrite (NO_2), is toxic to most plants. Nitrogen levels are extremely difficult to measure because:

(a) it is highly soluble and will be leached out quickly. Readings for fertilizer applications are useless
(b) nitrogen is continually being mineralized (recycled) from soil organic matter.

Nitrogen should therefore be applied as little as possible as often as possible. Only about 75 per cent of nitrate applied is used by plants. The rest is lost through leaching and runoff, often entering the water supply. For the purpose of fertilizer applications therefore, nitrogen levels are 'estimated' based upon:

(a) previous plants grown
(b) summer rainfall
(c) soil texture.

Fertilizers may be used in one of four different ways to supplement soil reserves: base dressing, top dressing, foliar feed or liquid feed (see fertilizer exercises). Some fertilizers are more suited to specific applications than others due to their differing rates of nutrient release characteristics. In this exercise the release rate of nitrogen from organic fertilizer is compared with an inorganic fertilizer.

Aim

To appreciate differences in the release rate of nitrogen from fertilizers.

Apparatus

Funnels	Retort stand	Clamps
Cotton wool	Measuring cylinder	Distilled water
Nitrate test strips	Beakers	Balance
Sandy soil	Ammonium nitrate	Hoof and horn

Method

1. Weigh out 1 g of each fertilizer and mix with 100 ml of distilled water.
2. Record the nitrate content of the solution using the test strips.

3. Record your result in the table provided.
4. Place a small piece of cotton wool in a funnel and position in a retort stand above a beaker.
5. Fill the funnel with a small amount of sandy soil.
6. Repeat steps 4 and 5 to make a second set.
7. Weigh out 5 g of inorganic fertilizer (ammonium nitrate) and broadcast on the top of one soil.
8. Weigh out 5 g of organic fertilizer (hoof and horn) and broadcast on the top of the other soil
9. Slowly add 25 ml of distilled water, in increments until the first leachate is collected.
10. Using the Merkoquant nitrate test strip record the nitrate content of the leachate.
11. Continue to add further 25 ml additions, until an additional 200 ml has been added, discarding the leachate between additions.
12. Record the resulting leachate nitrate content.

Results

Record your results in the tables provided.

Fertilizer solution	Nitrate content (mg/l)
1% w/v ammonium nitrate	
1% w/v hoof and horn	

Soil treatment	Leachate 1: nitrate content (mg/l)	Leachate 2: nitrate content (mg/l)	% change
Soil + ammonium nitrate			
Soil + hoof and horn			

Conclusions

1. Which fertilizer contains the highest nutrient content?
2. Which fertilizer is more soluble?
3. Which fertilizer treatment would be result in more nitrate leaching?
4. State an application use for the fertilizers with reference to the release rate of nitrogen, giving reasons for your answers (e.g. top dressing, base dressing etc.).
   (a) sulphate of ammonia
   (b) hoof and horn.
5. How should nitrogen be applied as a fertilizer?
6. Explain the role of nitrogen in plant growth and development.
7. Describe the typical symptoms of nitrogen deficiency in a named plant.

Answers

Exercise 18.1. Plant nutition

Results

1.   A plant nutrient is an element essential for life, without which the plant will die.

2.   16.

3.   A major nutrient is an element required in large quantities.

4.   The major elements are N, P, K, Ca, Mg, and S.

5.   A trace element is a nutrient required in small quantities which are toxic in larger quantities.

6.   The trace elements are Fe, Zn, Mn, Cu, B, Mo and Cl.

7.   Trace elements applied in large quantities are toxic to plants and will result in death.

8.   (a) macro nutrients
      (b) minor nutrients.

9.   Carbon, hydrogen and oxygen.

Conclusions

1.   The essential nutrients in sufficient quantity and in balanced proportions.

2.   An absence of any plant nutrient will result in plant death.

3.   Inadequate supply of nutrients will result in abnormal growth and deficiency symptoms showing in the leaves and growth patterns.

4.   Although Na, Si and Co are found to benefit plant growth, plants can grow without them and they are therefore not essential and not nutrients.

Exercise 18.2. Tomato plant deficiency symptoms

Results

Deficiency symptoms as those in Table 18.1 should be described. Distilled water and tap water show multiple deficiency symptoms since most of the nutrition has come from the cotyledon food store.

Conclusions

1.   Distilled water is the purest form of water containing no nutrients; it was used as a control to see how the seedlings grew.

2.   Tap water contains an inadequate supply of nutrients and will not result in healthy plant growth. Cut flowers should be fed a proprietary flower feed.

3.   Nitrogen showing stunted growth and pale leaves.

4.   (a) nitrogen deficiency
      (b) magnesium deficiency
      (c) potassium deficiency
      (d) phosphorus deficiency.

Exercise 18.3. Function of plant nutrient

Results

(a)   N
(b)   C, H
(c)   O
(d)   K
(e)   Mg
(f)   Ca
(g)   S
(h)   Trace elements
(i)   P

Exercise 18.4. Major nutrient roles

Results

1.   Nitrogen

2.   Magnesium

3.   Potassium

4.   Phosphorus

Conclusions

Major elements are nutrients required in large quantities. Minor elements are nutrients required in small quantities which can be toxic if supplied in large quantities.

Exercise 18.5. Nitrate fertilizers

Results

Ammonium nitrate will give higher nitrate results than hoof and horn because 50 per cent is immediately soluble. Hoof and horn is organic and releases nitrate slowly.

Conclusions

1.   Ammonium nitrate 34.5 per cent N.

2.   Ammonium nitrate, 50 per cent immediately soluble, 50 per cent slowly soluble.

3.   Ammonium nitrate.

4.   (a) sulphate of ammonia as a top dressing
      (b) hoof and horn as a base dressing before planting.

5.   Nitrogen should be applied little and often to prevent leaching.

6.   Nitrogen is essential for vegetative growth, protein and chlorophyll formation.

7.   Nitrogen deficiency can be recognized by stunted growth, light green to pale yellow leaves starting at the leaf tip, followed by death of the older leaves if the deficiency is great. In severe cases the leaves turn yellowish-red along their veins and die off quickly. Flowering is both delayed and reduced.

Chapter 19   Fertilizers

Key facts

1. Fertilizer bags always state content as the per cent N-P-K present (e.g. 16-10-12).
2. Eutrophication is the main environmental hazard linked to fertilizer use.
3. Nitrogen is the fertilizer required in largest quantity.
4. Fertilizers may be powdered, granulated or prilled.
5. Fertilizers may be straight, compound or blended.
6. Nutrient release rates may be fast, slow or controlled release.

Background

Fertilizers are a method of applying nutrients to the soil to enhance plant growth. Fertilizers can be natural (organic) or artificial (man-made). The amount of fertilizer added is based upon existing soil reserves of available nutrients and the intended plants to be grown. The quantity added should be sufficient to meet plant growth, without wastage (e.g. due to leaching). This requirement affects the type of fertilizer used (e.g. granular, prill, slow release), and the timing of applications (e.g. nitrogen, as little as possible, as often as possible). The primary concern to the environment from fertilizers is eutrophication. This is caused by excessive nutrient levels, particularly nitrogen and phosphorus leaching into watercourses such as lakes, ponds and streams. Eutrophication causes algal blooms and weed growth that lowers dissolved oxygen levels (hypoxic conditions) and greatly reduces water quality for fish and other animals.

Application of fertilizers

Fertilizers can be applied in one of four different ways:

Base dressing:	mixed into growth media, usually before planting
Top dressing:	broadcast onto the soil/growth media surface
Foliar feed:	normally to correct deficiencies sprayed onto leaves
Liquid feed:	in hydroponics production systems normally supplied direct to the roots.

Generally, fertilizers are used to supply nitrogen (N), phosphorus (P), potassium (K) or combinations of these. In addition, magnesium (Mg) is often needed together with trace elements. A range of fertilizers used to supply N, P, K and Mg and is presented in Table 19.1.

Nitrogen

Most plants absorb nitrogen as nitrate which is extremely soluble, making leaching a problem. The quantities of leached nitrates appearing in water supplies has often exceeded the EEC limit of 50 mg/l. Nitrogen should be added as little as possible, as often as possible, to avoid this. Many people now add nitrogen in a slow-release form. Additional nitrogen fertilizers include nitrate of soda (16%N), nitrochalk (21%N), urea (45%N), urea formaldehyde (40%N), gold N (36%), nitroform (38%), among others.

Phosphorus

Phosphorus is extremely insoluble. At pH 6.5 it is 70% insoluble, and in acid and alkaline conditions 100% insoluble. Phosphorus is bound tightly to clay particles but can be washed out of growth media easily.

Table 19.1 Nutrient content of some common straight fertilizers

Fertilizer name	Nutrient content	Solubility	Release rate	Form	Characteristics	Uses
Nitrogen						
Ammonium nitrate 33.5–34.5% N, e.g. Nitram. This is the commonest N fertilizer in the world	33.5/34.5-0-0	50% is immediately soluble and 50% slowly soluble	Very fast	Inorganic	Tends to make soils acid so needs extra lime. Useful early in season for overwintered crops	Top dressing Liquid feed Base dressing
Sulphate of ammonia 21%N. Most favoured amateur N fertilizer. Cheap	21-0-0	Soluble	Fast	Inorganic	Has a greater acidifying action than all other nitrogen fertilizers and is less efficient than nitrate forms	Top dressing Liquid feed Base dressing
Calcium nitrate 15% N. Does not acidify the soil. More expensive than other N sources	15.5-0-0	Soluble	Fast	Inorganic	Mainly used on horticultural plants were acid soils are not desired	Top dressing Liquid feed Base dressing
Dried blood 10–13% N. Quick-acting organic fertilizer	12-13-0	Partly soluble	Fairly to very fast	Organic	Especially suited for liquid feeding	Liquid feed Top dressing
Hoof and horn 13% N. Ground up hoofs and horns of cattle	13-14-0	Insoluble slow release	slow to fairly fast	Organic	Rate of release depends upon size of particles	Base dressing Top dressing
Phosphorus						
Mono ammonium phosphate. Best known of the water-soluble forms	12-25-0	Soluble	Fast	Inorganic	Made from Rock phosphate, ammonia and phosphoric acid	Liquid feed Top dressing
Superphosphate. Also known as single superphosphate	0-18/2-0	Soluble	Fast	Inorganic	Contains calcium sulphate which is acidic and can raise conductivity	Top dressing Base dressing
Triple superphosphate. Another water-soluble straight P source	0-47-0	Soluble	Fast	Inorganic	Phosphorus rock mixed with acid. Contains no impurities. Neither form of superphosphate is suited for liquid feeding	Top dressing Base dressing
Bone meal (should be sterilized)	0-22-0	Slowly soluble	Slow	Organic	Expensive forms of P that contain no advantages over the inorganic forms. Also contains a little N	Base dressing
Steamed bone meal/flour (should be sterilized)	0-28-0	Slowly soluble	Slow	Organic	Both bone meals are in little demand, except for market garden and amateur use	Base dressing

(Continued)

Table 19.1 (Continued)

Fertilizer name	Nutrient content	Solubility	Release rate	Form	Characteristics	Uses
Potassium						
Sulphate of potash	0-0-50	Soluble	Fast	Inorganic	Manufactured to form a dense powder that does not cake easily	Base dressing Top dressing Liquid feed
Potassium nitrate	13-0-44	Soluble	Fast	Inorganic	Also known as Saltpetre. Expensive. Mainly used as a liquid feed	Liquid feed Top dressing
Muriate of potash (potassium chloride)	0-0-60	Soluble	Fast	Inorganic	Small dry crystals that do not cake easily. Cheap, often used on fruit, but chloride content can raise conductivity and damage soft fruits	Foliar feed Base dressing
Rock potash	0-0-50	Slowly soluble	Slow	Organic	Difficult to obtain the natural mineral and rarely used. Needs to be ground	Base dressing
Magnesium						
Epsom salts (magnesium sulphate)	0-0-0-10	Soluble	Fast	Inorganic	Useful for foliar spray or liquid drench. This is an expensive form of Mg	Foliar feed Liquid feed Top dressing Base dressing
Kieserite (another form of magnesium sulphate)	0-0-0-17	Slowly soluble	Very slow	Inorganic	Mg becomes available to plants in the first season, and then for the following next 2 or 3 seasons	Base dressing
Magnesian Limestone (dolamitic limestone)	0-0-0-8/12	Slowly soluble	Slow	Inorganic	Used when magnesium levels are low and acidity needs to be corrected	Base dressing

Fungi associations help plants to take up P. Where possible water-soluble fertilizers should be used. Manufacturers will always state on the bag what percentage of the mix is water and acid soluble. Phosphorus content is stated as phosphorus pentoxide (P_2O_5). Additional phosphorus fertilizers include rock phosphate (25–40%), basic slag (12–18%), di-ammonium phosphate (54%) and blood, fish and bone (3.5-8-0.5 variable).

Potassium

Potassium is generally readily soluble in water, but is not easily leached. Luxury supply of potassium should be avoided. Potassium is associated with the health and efficient functioning of plants and is generally applied in proportion to nitrogen. The potassium content of fertilizers is declared potash (K_2O), otherwise called potassium oxide. Additional potassium fertilizers include Chilean Potash nitrate (10–15%), Kainit (14–30%), and farmyard manure (40%).

Magnesium

Magnesium is required in about 10% of the plant's N and K requirements, roughly equal to P, and hundred times greater than trace elements. Most soils have adequate reserves of Mg. Magnesium content is stated as % Mg. Additional Mg fertilizers are rare. Farmyard manure has been found to contain 1%, basic slag 0.5%. In the UK, rainfall provides 2 kg/ha, for each 100 mm of rain.

Fertilizer bag content

Fertilizers list N-P-K-Mg as a percentage of the weight of the fertilizer, e.g. Osmocote (18-11-10), Ficote 140 (16-10-10). Thus, Osmocote is 18 per cent nitrogen, 11 per cent phosphorus and 10 per cent potassium. Similarly, Ficote 140 is 16 per cent nitrogen, 10 per cent phosphorus and 10 per cent potassium. The remaining material is impurities and a carrier material such as sand, used to aid spreading the fertilizer.

Exercise 19.1

Fertilizer nutrient content

Aim

To increase knowledge of general N, P, K and Mg fertilizer frequently used in horticulture.

Apparatus

> background text
> Table 19.1

Useful websites

www.efma.org

www.environment-agency.gov.uk

www.defra.gov.uk

Method

Answer the list of questions provided.

Results

1. State the percentage nitrogen content of ammonium nitrate.

2. List an organic nitrogen fertilizer suitable for use as a base dressing prior to planting shrubs.
3. Explain the difficulties in applying phosphorus as a fertilizer.
4. State the percentage phosphorus content, and the uses, of triple superphosphate.
5. Sulphate of potash is a common garden centre retailed fertilizer. State its nutrient content.
6. Explain why rock potash is unsuitable for use as a foliar feed.
7. Epsom salts is a very common form of magnesium fertilizer. What is its chemical name and nutrient content?
8. Select a slow-release magnesium fertilizer suitable for use as a base dressing.

Conclusions

1. Explain the difference between a 'base dressing' and a 'top dressing'.
2. Explain the difference between a 'liquid feed' and a 'foliar feed'.
3. A rose fertilizer has a nutrient content of 5-6-12. Explain what this means.

Exercise 19.2

Structure of fertilizers

Background

There are three categories of solid fertilizers available to horticulturalists. These are:

- **Powdered fertilizer** in the form of powder or dust (e.g. hoof and horn).
- **Granulated fertilizers** which comprise non-spherical grains from 1 to 4 mm in diameter. They tend to be rough in texture, and irregular in both size and shape. Granular materials may be straight N, P, K, or any proportion of each. Straight N fertilizer is particularly common to this group. The materials are hard and have a higher density than powdered, therefore making them easier to spread.
- **Prilled fertilizer** comprises spherical shaped balls of uniform size (approximately 2 mm), but can vary from less than 1 mm up to 3.5 mm. They are round, smooth and very free flowing. They can be coated to make them harder and less susceptible to moisture absorption. Straight nitrogen is the most common prilled fertilizer, either as ammonium nitrate or as urea.

Aim

To identify correctly the structure of fertilizers.

Apparatus

Weigh boats	Ammonium nitrate
Steamed bone flour	Copper sulphate
Kieserite	Ficote
Table 19.1	

Useful websites

www.fertilizer.org/ifa

www.efma.org

Method

1. Decant a small amount of each fertilizer into a weigh boat.
2. Observe the size, shape and texture of the fertilizers and decide whether they are powdered, granular or prilled.
3. Reading from Table 19.1 or other sources, state the type and percentage nutrient supplied.

Results

Enter your results in the table provided.

Fertilizer	Structure	Nutrient supplied	% of nutrient in fertilizer
Ammonium nitrate			
Steamed bone flower			
Copper sulphate			
Kieserite			
Ficote			

Conclusion

1. Explain, based upon your observations, which fertilizers may be used for:
   (a) base dressing
   (b) top dressing.
2. State two other methods of applying fertilizers.

Exercise 19.3

Fertilizer spreading

Background

The structure of fertilizers is important, largely due to the ease of spreading. This in turn is influenced by:

(a) particle size (prills, granules, etc.)
(b) density
(c) hardness of material.

These properties combined can add up to the difference between throwing a ping-pong ball or a golf ball. Materials with larger particle sizes and higher densities will be better for spreading. Problems from poor spreading include 'streaking', the dark- and light-green stripes in a lawn.

Material hardness

This test can also be used to assess both the age and solubility of fertilizers. Older fertilizers crumble more easily as do more soluble ones. Try to crush some fertilizer with the back of your finger nail. If the fertilizer:

- crumbles, the product is poor
- crumbles with difficulty, the product is good
- impossible to crush, the product is very good.

Aim

To assess the age, solubility and ease of spreading of fertilizers.

Apparatus

Weigh boats	Ammonium nitrate
Steamed bone flour	Copper sulphate
Kieserite	Ficote

Method

Grade the fertilizers using the material hardness test.

Results

Enter your results in the table provided.

Fertilizer	Material hardness	Interpretation
Ammonium nitrate		
Steamed bone flower		
Copper sulphate		
Kieserite		
Ficote		

Conclusions

1. Which materials would be the easiest to spread?
2. Which materials are the oldest stock?
3. Which materials are probably the most soluble?
4. List six major nutrients required by plants for healthy growth.
5. Name four fertilizers commonly used as top dressings.

Exercise 19.4

Types of fertilizer

Background

There are three basic types of fertilizer formulations available:

1. **Straight fertilizer** – These contain only one of the major plant nutrients, e.g. ammonium nitrate
2. **Compound** – These fertilizers contain two or more nutrients bonded together, e.g. Growmore 7-7-7.
3. **Mixed/blended** – Produced by mixing two or more straight fertilizers together, e.g. John Innes Base Fertilizer, not a common method.

Aim

To identify types of fertilizers.

Apparatus

Weigh boat	Ammonium nitrate
Urea	Potassium nitrate
Ammonia sulphate	Sulphate of potash
Hoof and horn	Table 19.1

Useful websites

www.efma.org

www.fertilizer.org

Method

1. Decant a small quantity of each fertilizer into a weigh boat.
2. Examine each material with its packaging and decide whether each fertilizer is straight, copmound or mixed/blended.

Results

Enter your results in the table provided.

Fertilizer	Straight, compound or blended	Nutrient(s)	% supplied
Ammonium nitrate			
Urea			
Potassium nitrate			
Ammonium sulphate			
Sulphate of potash			
Hoof and horn			

Conclusions

1. Which fertilizer might you use as a top-dressing in summer for lawns?
2. Which fertilizer might you use as a base dressing for a shrubery in spring?
3. State what is meant by:
   (a) straight fertilizer
   (b) compound fertilizer.
4. State a fertilizer that may be used as a foliar feed to correct nitrogen deficiency.

Speed and mode of nutrient release

Background

Fertilizers may also be classified according to how fast or slow they release their nutrients into the soil or growth medium.

Fast release

With all these fertilizers the nutrients are available immediately, and can be taken up by the plant, giving an immediate response. This group includes those that are readily soluble. Often horticulturalists rank these according to the speed of reaction of the plant to the fertilizer. However, fast-release fertilizers can leach away quickly and if applied too heavily, cause damage to the plant.

Slow release/controlled release

These may be tablets, granules, powders or even stick fertilizers whose nutrition is not immediately soluble. They release their nutrients slowly over an extended period of time based upon their rate of decomposition/ mineralization. Three such groups can be identified:

- **Organic fertilizers**. Nutrients become available by the slow breakdown of organic materials and the mineralization resulting from the actions of microorganisms, e.g. bone meal. Organic fertilizers take a long time to release nutrients and are useful as a base dressing prior to planting. To be effective the soil needs to be moist and warm enough to encourage soil microorganism activity.
- **Slowly soluble fertilizers** (S-R-F). These break down slowly as water absorbs and bacteria degrade into them over longer periods of time than those fertilizers that are immediately soluble, e.g. urea formaldehyde. They have the advantage that they do not cause rapid, weak growth, leaching and can be applied less often than fast-release fertilizers.
- **Controlled release** (C-R-F). These are resin- or polymer-coated fertilizers. Their skins gradually decompose to release nutrients by the action of water and temperature.

New products are appearing regularly. The most widely used C-R-F are Osmocote and Ficote. Rates of release vary from 8 weeks to 18 months. Generally, however, they may be divided into two:

(a) single season (8–9 months), e.g. Ficote 70 (16-10-10)
(b) extended season (12–42 months), e.g. Ficote 140 (16-10-10) and Ficote 360 (16-10-10).

When the resin coats absorb water their contents are released slowly, through sub-microscopic pores. The major factor controlling the rate of flow of nutrients through the skin is temperature. As the temperature increases so does the porosity of the coat. In this way the release of nutrients matches the plant demand. To prevent excess buildup of nutrients in the growth media, the porosity decreases, cutting off the supply of nutrients, if the temperature rises too high. In the example of Ficote the number preceding the nutrient content refers to the number of days over which the nutrients are released, above the critical environmental conditions. For example, Ficote 70 releases nutrients over a 70-day period. Ficote 360 releases nutrients across a 360-day period. However, the release period may be spread across 12–42 months and include periods during the winter when no nutrients are being released.

Aim

To recognize modes of fertilizer release.

Apparatus

Weigh boats	Osmocote
Ficote	Ammonium nitrate hoof and horn
Dried blood	Potassium nitrate
Mono ammonium phosphate	Beaker
Water	Spatula
Table 19.1	

Method

1. Decant some of each fertilizer into a weigh boat.
2. Conduct a material hardness test (Exercise 19.3) and note the fertilizer response when mixed in the beaker with a little water.
3. Decide whether each fertilizer is fast, organic, slowly soluble or controlled release.

Results

Enter your results in the table provided.

Fertilizer	Speed of nutrient release (fast, organic, slow or controlled)
Osmocote	
Ficote	
Ammonium nitrate	
Hoof and horn	
Dried blood	
Potassium nitrate	
Mono ammonium phosphate	

Conclusions

1. Based upon your observations, state which of the above listed fertilizers may be used for:
   (a) base dressing
   (b) top dressing
   (c) foliar feed
   (d) liquid feed.
2. Describe the role of controlled-release fertilizers and fast-release fertilizers in horticulture.
3. Explain why hoof and horn fertilizer may be ineffective when used in spring as a nitrogen top dressing.
4. Name a fertilizer suitable for each of the following uses:
   (a) an acidifying nitrogen fertilizer for turf or field use
   (b) a fertilizer that will supply phosphate for liquid feeds
   (c) a fertilizer for liquid feeding that is also a liming material
   (d) a magnesium fertilizer that is also a liming material.
5. State two examples of slow-release fertilizers suitable for use with container-grown plants.

Exercise 19.6

Fertilizer recommendations and calculations

Background

Fertilizer recommendations are based upon soil analysis and the requirements of the plant to be grown, making an allowance for nutrient residues left from the previous plants grown. All fertilizer recommendations, such as from analysis laboratories, Agricultural Development Advisory Service (ADAS), reference books, amateur journals (e.g. *Gardening Which?*) and in manufacturers' data sheets, are reported in kg/ha, e.g:

N = 220 kg/ha (N)
P = 250 kg/ha (P_2O_5)
K = 300 kg/ha (K_2O)

This may be applied either as straight fertilizers or by choosing a compound fertilizer to give as near as possible the correct amount. In this case the priority is to get the nitrogen supply right. Slight variations in the rate of phosphorus and potassium will have less effect.

Aim

The end user will have to interpret this recommendation according to the fertilizers they have in stock. You must therefore be able to translate the recommendation into its equivalent for the fertilizer that you have available. This can be achieved by using the following formulae:

$$\text{Amount of fertilizer required in kg/ha} = \frac{\text{Nutritional requirement advised (in kg/ha)}}{\text{\% nutrient concentration in alternative fertilizer}} \times 100$$

Method

Work through the two example calculations below, and then answer the questions. You need not answer every question; when you have confidently mastered this skill, proceed to Exercise 19.7.

Example 1

ADAS recommend applying 36 kg/ha phosphorus pentoxide (P_2O_5). The only fertilizer available is single superphosphate (18% P_2O_5).

Using the above formulae:

$$\text{Single superphosphate required in kg/ha} = \frac{36\ \text{kg/ha}}{18\%} \times 100$$
$$= 2 \times 100 = 200\ \text{kg/ha}$$

To convert kg/ha to g/m^2, divide by 10, e.g. 200 kg/ha = 20 g/m^2 of single superphosphate.

Example 2

ADAS recommend 36 kg/ha phosphorus pentoxide and you have triple superphosphate (TSP), (46% P_2O_5) available:

$$\text{Amount of TSP required} = \frac{36\ \text{kg/ha}}{46\%} \times 100$$
$$= 0.78 \times 100$$
$$= 78\ \text{kg/ha triple superphosphate}$$
$$= \frac{78}{10} = 7.8\ \text{g/m}^2$$

Results

1. Calculate the amount of mono ammonium phosphate (25% P_2O_5) required to provide a top dressing recommended in *Gardening Which?*, of 36 kg/ha phosphorus pentoxide.
2. Calculate the amount (in kg/ha and g/m^2), that would be required to provide a base dressing of potash (K_2O) of 150 kg/ha, using muriate of potash (60% K_2O).

3. A handbook for lawns indicates that the lawns require a top dressing of 100 kg/ha, nitrogen (N). How much ammonium nitrate (33.5% N), should you use?

4. As a result of a soil analysis report, the *Elaeagnus pungens* stock plant beds require a top dressing of 50 kg/ha N. How much sulphate of ammonia (21% N) should you use?

5. Analysis results show that the soil in the all-year-round *Chrysanthemum* house is becoming unacceptably acidic, due to the previous fertilizer regime. The nitrogen recommendation is for 200 kg/ha N. You have checked the stores and have both sulphate of ammonia (21% N), and calcium nitrate (15% N), available. Which one should you use? Why? And at what application rate?

6. The lettuce plants require a top dressing of 150 kg/ha N. The manager has decided to grow them organically because he believes that this will yield him a premium price. How much dried blood (12% N) should be used?

7. A local grower has decided to establish an organically grown Pick-Your-Own strawberry unit. How much hoof and horn (13% N) should be used to meet the recommended base dressing rate of 75 kg/ha?

8. How much single superphosphate (18% P_2O_5), is required to complete a turf seed bed preparation rate specified at 75 kg/ha phosphorus pentoxide?

9. Calculate the rate at which sulphate of potash (50% K_2O), should be applied to supply:
   (a) 50 kg/ha
   (b) 85 kg/ha.

10. The tree stock production unit is to be moved. The soil has been sampled and the analysis results indicate the following requirements:
    N: 100 kg/ha (N)
    P: 25 kg/ha (P_2O_5)
    K: 200 kg/ha (K_2O)
    Mg: 75 kg/ha (Mg)
    How much of the following fertilizers would be needed ?
    (a) Ammonium nitrate (33.5% N)
    (b) Superphosphate (20% P_2O_5)
    (c) Sulphate of potash (50% K_2O)
    (d) Kieserite (17% Mg).

11. The annual top dressing requirement of the rose beds as reported in the *Daily Mail* gardening section is as follows:
    N: 100 kg/ha (N)
    P: 37 kg/ha (P_2O_5)
    K: 65 kg/ha (K_2O)
    Mg: 41 kg/ha (Mg).
    The following fertilizers are available for use. How much of each is required?
    (a) Ammonium nitrate (33.5% N)
    (b) Triple superphosphate (46% P_2O_5)
    (c) Sulphate of potash (50 % K_2O)
    (d) Kieserite (17% Mg).

12. 50 kg/ha of magnesium is required for a seed bed preparation.
    (a) Which magnesium fertilizer would you use?
    (b) At which rate would you apply it?

13. After sampling and analysing a nursery soil, it is found that fertilizer is required at the following application rates:

N: 180 kg/ha

P_2O_5: 90 kg/ha

K_2O: 90 kg/ha

Choose a suitable fertilizer from the list to use and state its rate of application:

Fertilizer A: 8:16:16

Fertilizer B: 20:10:10

Fertilizer C: 24:18:18

Conclusions

1. Explain why potassium fertilizers are easy to apply as a foliar feed where as phosphorus ones are not.
2. State how magnesium deficiency may be overcome.
3. One 50 kg bag of fertilizer contains 15% nitrogen, 15% phosphorus (P_2O_5), and 20% potassium (K_2O). How many bags of this fertilizer are required to supply 75 kg N, 75 kg P_2O_5, and 100 kg K_2O per hectare?
4. Name four fertilizers commonly used as base dressings.
5. State the role of nitrogen in the plant and the precautions to be taken when using nitrogenous fertilizers.
6. Which one of the following is a phosphorus fertilizer?
   (a) Basic slag
   (b) Nitrate of potash
   (c) Hoof and horn
   (d) Urea.

Exercise 19.7

Area and amount of fertilizer required

Background

Most horticulturalists need to know how much fertilizer to get out of the fertilizer store to complete the application recommended without any waste, either in terms of chemical, or time spent returning to the store to get additional quantities of fertilizer or returning surplus.

For this we need to know the 'area' of land that is to be treated with fertilizer. This is calculated simply by multiplying the length by the width of the area to be treated. For example:

(a) Tree production unit is 20 m by 50 m

Area = 20 m $\times$ 50 m = 1000 m^2

(b) *Alstroemeria* beds measure 2 m by 30 m.

Area = 2 m $\times$ 30 m = 60 m^2

Aim

To increase ability in calculating areas for fertilizer application rates.

Method

Calculate the areas of the following units.

Results

1. AYR chrysanthemum beds measuring 4 m by 30 m.
2. Polytunnel measuring 15 m by 20 m.
3. Efford sand beds measuring 8 m by 15 m.
4. Pot chrysanthemum benches measuring 25 m by 40 m.
5. Ericacae stock beds measuring 1.5 m by 8 m.

Conclusions

Explain the relationship between the area of land management and fertilizer recommendations.

Exercise 19.8

Total fertilizer quantity measurements

Background

Using this area measurement (m^2) (see Exercise 19.7), and the fertilizer application rate (g/m^2) (see Exercise 19.6), we can now calculate the total quantity of fertilizer required using the following formulae:

Total quantity (Q) of fertilizer required = Area (m^2) $\times$ Application rate (g/m^2)

Example

A base dressing fertilizer has a rate of application of 75 kg/ha. How much fertilizer would be needed for a seed bed measuring 20 m $\times$ 50 m?

$$
\begin{aligned}
\text{Step A:} \quad \text{Area} \quad &= 20\,m \times 50\,m = 1000\,m^2 \\
\text{Step B:} \quad \text{Quantity Q} &= 1000\,m^2 \times 7.5\,g/m^2 \\
&= 7500\,g \\
&= 7.5\,kg
\end{aligned}
$$

Method

Calculate the total quantity of fertilizer required in the following situations.

Results

1. A base dressing of 75 kg/ha N for seed bed production of celery in a polytunnel measuring 20 m by 50 m. How much fertilizer is needed?
2. A base dressing of 150 kg/ha is recommended for a glass house measuring 40 m by 50 m. How much fertilizer is required?
3. Phosphorus pentoxide is required at 75 kg/ha. How much is needed for an *Alstroemeria* bed measuring 2 m by 15 m, using:
   (a) single superphosphate (18% P_2O_5)
   (b) triple superphosphate (46% P_2O_5)?
4. A garden border measures 6 m by 20 m. Soil analysis recommendations state a base dressing of:
   (a) Ammonium nitrate at 70 g/m^2
   (b) Triple superphosphate at 150 g/m^2
   (c) Sulphate of potash at 350 g/m^2
   (d) Kieserite at 240 g/m^2
   How much of each of these fertilizers is needed?

Conclusions

(Covering all exercises in this chapter.)

1. State the role of nitrogen, phosphorus, potassium and magnesium in plant growth. Describe the symptoms of their deficiency in a named plant.
2. Explain the purpose of using fertilizers in horticulture
3. Distinguish between the following terms:
    (a) base dressing
    (b) top dressing
    (c) foliar feed
    (d) liquid feed.
4. State the chemical units that fertilizer recommendations quoted in for:
    (a) nitrogen
    (b) phosphorus
    (c) potassium
    (d) magnesium
5. Describe how the method of nutrient release varies between fast and slow.
6. Explain the principle of nutrient release from controlled-release fertilizers.
7. Name two straight fertilizers which supply phosphorus and potassium.
8. Describe two advantages and two disadvantages of using organic fertilizers compared to inorganic fertilizers.
9. In which one of the following soils is potassium deficiency most likely to occur?
    (a) badly drained
    (b) heavy texture
    (c) sandy textured
    (d) well drained.

Answers

Exercise 19.1. Fertilizer nutrient content

Results

1. 33.5–34.5 per cent.

2. Hoof and horn.

3. Phosphorus is extremely insoluble and tightly held by clay particles.

4. TSP is 47 per cent phosphorus and may be used as both a top and base dressing.

5. 0-0-50.

6. Because it is only slowly soluble and will not mix easily with water.

7. Magnesium sulphate 0-0-0-10.

8. Kieserite.

Conclusions

1. A base dressing is mixed into the growth medium before planting. A top dressing is broadcast onto the surface after planting.

2. A liquid feed is applied as an irrigation to the growth medium. A foliar feed is sprayed onto the leaves.

3. 5 per cent nitrogen, 6 per cent phosphorus and 12 per cent potassium.

Exercise 19.2. Structure of fertilizers

Results

Ammonium nitrate	granular	N	33.5–34.5%
Steamed bone flour	powder	P	28%
Copper sulphate	granular	Cu	n/a
Kieserite	powder	Mg	17%
Ficote	prilled	N-P-K	16–10–10

Conclusions

1. (a) Granular and prilled are suitable for base dressings because they are longer lasting in the soil e.g. Ficote
   (b) Powdered fertilizers are suitable for top dressing after planting because they are more quickly decomposed to release their nutrients e.g. kieserite.

2. Liquid feed and foliar feed.

Exercise 19.3. Fertilizer spreading

Results

Ammonium nitrate	crumbles with difficulty	product is good
Steamed bone flower	crumbles	product is poor
Copper sulphate	crumbles	product is poor
Kieserite	crumbles with difficulty	product is good
Ficote	impossible to crush	product is very good

Conclusions

1. Ficote. It is of uniform size and weight and easily broadcast.

2. Those that crumble easiest are either the oldest stock or most soluble.

3. Those that crumble the easiest, e.g. copper sulphate.

4. Choice from: C, H, O, N, P, K, Mg, Ca and S.

5. Any four stated as suitable for top dressing from Table 19.1, e.g. ammonium nitrate, sulphate of ammonia, calcium nitrate, dried blood, hoof and horn, etc.

Exercise 19.4. Types of fertilizer

Results

Ammonium nitrate	straight	N	33.5–34.5%
Urea	straight	N	45%
Potassium nitrate	compound	N, K	13-0-44
Ammonium sulphate	straight	N	21%
Sulphate of potash	straight	K	50%
Hoof and horn	blended	N, P	13–14–0

Conclusions

1. A nitrogen fertilizer for vegetative growth that is easily soluble, e.g. ammonium nitrate or urea.

2. A compound fertilizer containing a mixture of nutrients in a slowly soluble form, e.g. potassium nitrate, hoof and horn or Ficote.

3. (a) A straight fertilizer contains only one nutrient, e.g. ammonium nitrate (34.5% N).

   (b) A compound fertilizer contains two or more nutrients chemically bonded together, e.g. potassium nitrate (13–0–44).

4. Ammonium nitrate since it is 50% immediately soluble and 50% slowly soluble.

Exercise 19.5. Speed and mode of nutrient release

Results

Osmocote	Controlled release
Ficote	Controlled release
Ammonium nitrate	Fast release
Hoof and horn	Organic
Dried blood	Fast organic
Potassium nitrate	Fast
Mono ammonium phosphate	Fast

Conclusions

1. (a) Osmocote, Ficote or hoof and horn;
   (b) ammonium nitrate, dried blood, or potassiun nitrate;
   (c) Monoammonium phosphate, ammonium nitrate or potassium nitrate;
   (d) Monoammonium phosphate, ammonium nitrate or potassium nitrate.

2. Controlled-release fertilizers are especially useful for base dressings or mixing in container compost because they release their nutrition over extended seasons and help to provide the customer with a better quality product. Fast-release fertilizers are better as top dressings broadcast onto the surface of the soil providing a periodic boast to nutrition.

Similarly fast release enables liquid and foliar feeding to be immediately beneficial to the plant.

3. Hoof and horn is an organic fertilizer that releases nutrients slowly over several seasons. In spring nitrogen will be required immediately to stimulate growth and so a fast-release fertilizer should be selected.

4. Using Table 19.1 as a guide:
   (a) sulphate of ammonia or ammonium nitrate;
   (b) mono ammonium phosphate;
   (c) calcium nitrate;
   (d) magnesian limestone.

5. Osmocote and Ficote

Exercise 19.6. Fertilizer recommendations and calculations

Results

1. $36/25 \times 100 = 144\,kg/ha = 14\,g/m^2$

2. $150/60 \times 100 = 250\,kg/ha = 25\,g/m^2$

3. $100/33.5 \times 100 = 298\,kg/ha = 30\,g/m^2$

4. $50/21 \times 100 = 238\,kg/ha = 24\,g/m^2$

5. Use calcium nitrate because it is not acidic at:
    $200/15 \times 100 = 1300\,kg/ha = 130\,g/m^2$

6. $150/12 \times 100 = 1250\,kg/ha = 125\,g/m^2$

7. $75/13 \times 100 = 577\,kg/ha = 58\,g/m^2$

8. $75/18 \times 100 = 417\,kg/ha = 42\,g/m^2$

9. (a) $50/50 \times 100 = 100\,kg/ha = 10\,g/m^2$
    (b) $85/50 \times 100 = 170\,kg/ha = 17\,g/m^2$

10. (a) $100/33.5 \times 100 = 298\,kg/ha = 30\,g/m^2$
    (b) $25/20 \times 100 = 125\,kg/ha = 13\,g/m^2$
    (c) $200/50 \times 100 = 400\,kg/ha = 40\,g/m^2$
    (d) $75/17 \times 100 = 441\,kg/ha = 44\,g/m^2$

11. (a) $100/33.5 \times 100 = 298\,kg/ha = 30\,g/m^2$
    (b) $37/46 \times 100 = 80\,kg/ha = 8\,g/m^2$
    (c) $65/50 \times 100 = 130\,kg/ha = 13\,g/m^2$
    (d) $41/17 \times 100 = 241\,kg/ha = 24\,g/m^2$

12. (a) Kieserite, slow release over two seasons
    (b) $50/17 \times 100 = 294\,kg/ha = 29\,g/m^2$

13. Fertilizer B applied at:
    $180/20 \times 100 = 900\,kg/ha = 90\,g/m^2$

Conclusions

1. Potassium is readily soluble, phosphorus is not, even at pH 6.5 it is 70 per cent insoluble and therefore difficult to apply as a foliar feed.

2. Foliar application of magnesium sulphate.

3. First determine N and P; $75/15 \times 100 = 500\,kg/ha = 10$ bags. Next determine K; $100/20 \times 100 = 500\,kg/ha = 10$ bags. Thus 10 bags of fertilizer 15-15-20 are required.

4. Choice of appropriate four from Table 19.1.

5. Nitrogen is primarily for vegetative growth. It can be easily leached and should be applied little and often rather than in large doses.

6. Hoof and horn (13-14-0).

Exercise 19.7. Area and amount of fertilizer required

Results

1. $4 \times 30 = 120\,m^2$

2. $15 \times 20 = 300\,m^2$

3. $8 \times 15 = 120\,m^2$

4. $25 \times 40 = 1000\,m^2$

5. $1.5 \times 8 = 12\,m^2$

Conclusions

Fertilizer recommendations are normally given in kg/ha. This can be converted to g/m^2 by multiplying by one hundred.

Exercise 19.8. Total fertilizer quantity measurements

Results

1. $(20 \times 50) \times 7.5 = 7.5\,kg$

2. $(40 \times 50) \times 15 = 30\,kg$

3. (a) $75/18 \times 100 = 416\,kg/ha = 42\,g/m^2 \times (2 \times 15) = 1.2\,kg$
    (b) $75/46 \times 100 = 163\,kg/ha = 16\,g/m^2 \times (2 \times 15) = 0.5\,kg$

4. (a) $(6 \times 20) \times 70 = 8.4\,kg$
    (b) $(6 \times 20) \times 150 = 18\,kg$
    (c) $(6 \times 20) \times 350 = 42\,kg$
    (d) $(6 \times 20) \times 240 = 29\,kg$

Conclusions

1. Nitrogen: vegetative growth; stunted growth. Phosphorus: root growth: purple leaf and stem colouring.

Potassium: stem strength and resistance to disease: leaf yellowing starting at edges.

Magnesium: photosynthesis: leaf yellow blotches.

2. (a) Fertilizers are used to replace nutrition used by the plant each year. They may be applied as a base-dressing, top-dressing, foliar feed or liquid feed.

3. (a) Base dressing is a fertilizer mixed into the soil before planting.
   (b) A top dressing is a fertilizer broadcast onto the soil surface after planting.
   (c) A foliar feed is soluble fertilizer sprayed onto the leaves usually to correct deficiency symptoms.
   (d) A liquid feed is a soluble fertilizer applied as a drench to the root zone.

4. (a) N
   (b) P_2O_5

(c) K_2O
(d) Mg

5. Fast release refers to fertilizers where the nutrition is easily soluble. Slow release requires the nutrition to be releases after the action of water, temperature or microoganisms breaking down and mineralizing the fertilizer.

6. Controlled-release fertilizers (C-R-Fs) are resin- or polymer-coated. The skin is gradually decomposed to release nutrients by the action of water and temperature.

7. Mono ammonium phosphate and sulphate of potash respectively.

8. Organic fertilizers are better for the environment and mostly slow release.

9. (a) badly drained.

Part Three

Pest and disease

Chapter 20   Fungi

Key facts

1. Over 80 000 species of fungi have been classified.
2. Fungi are the cause of many plant diseases including *Botrytis*, Ccoral spot, damping-off, rusts and smuts.
3. Individual fungi are the largest life forms on the planet, often spreading over many acres within the soil.
4. Fungi that cause diseases are either saprophytes or parasites and enter plants through wounds or weak tissue.
5. Fungi are cryptogams, reproducing by spores.
6. Spore mapping is often the only way to correctly identify specific species of fungus.

Background

The non-green plants, such as bacteria and fungi, do not normally contain chlorophyll, and have to obtain their nourishment from other organisms. This type of plant is known as a **parasite**. They are a major cause of disease, infecting plants in all areas of horticulture. However, a knowledge of their biology can aid us in the timing of control measures.

Fungi are the 'decomposers' in the food chain. They help maintain soil fertility by recycling nutrients and decomposing organic matter. They secrete digestive enzymes onto the material which, once soluble, are absorbed into the fungi. They are cryptogamous organisms which, because they lack chlorophyll, feed on organic matter. (A cryptogam is a plant that reproduces by spores.) Fungi help clean the air, and provide food, materials and medicine.

There are over 80 000 different species of fungi that have been classified out of a total estimate of 1.5 million species. They are eukaryotic: their cells contain many organelles that are also present in plant and animal cells (see Chapter 1, Table 1.2). The largest is the puff ball and the smallest are single-celled organisms such as yeast. They may be saprophitic (feeding on dead and decaying matter), or parasitic (e.g. bracket fungi and athlete's foot, *Phytophthora*, *Pythium*, *Rhizoctonia*, *Fusarium* and *Botrytis*).

Saprophytes live on dead and decaying material and help to recycle nutrients including carbon and nitrogen.

Obligate saprophytes only use dead material.

Facultative saprophytes prefer dead matter, but can use living material.

Parasites derive their nutrition from living materials.

Obligate parasites can only use live food (e.g. powdery and downy mildew, rusts, smuts). If you clear the site of live food you will kill the parasite. Powdery mildew, like other obligate parasites, weaken the plant but do not kill it.

Facultative parasites prefer live food, but can survive on dead. For prevention, horticulturalists must destroy all dead plant material as well as living waste. For example, honey fungus can live on dead matter for over twenty years before moving onto live food. Hygiene is therefore essential – always remove as much dead material as possible.

Common fungi include the mould fungi which grow on stale bread and cheese. Other examples of fungi include mucor, coral spot rust fungus, mushrooms, toadstools, puff balls and bracket fungi.

Figure 20.1 shows a typical coral spot disease (*Nectria cinnabarina*). The saprophyte, coral-coloured mycelium decompose the dead stems and branches of a wide range of trees and shrubs, including fruit bushes and untreated fences.

Figure 20.1 Coral spot disease

Fungi can be classified into four groups (see Table 20.1).

Table 20.1 Fungus classification

Class	Example fungi and diseases
Primitive fungi (Phycomycetes) (algae-like fungi)	Club root (*Plasmodiophora brassicae*)
	Damping off (*Pythium, Phytophthora*)
	Downy mildew (*Plasmopara viticola, Perenospora, Bremia*)
	Potato blight (*Phytophthora infestans* and *parasitica*)
Intermediate fungi (Ascomycetes) (sac fungi, over 30 000 species)	Powdery mildew (*Erisyphe*)
	Leaf spots (*Cercospora, Pseudopeziza*)
	Stem rots (*Didymella, Nectria*)
	Wilts (*Fusarium, Verticillium*)
	Dutch elm disease (*Ceratocytis ulmi*)
	Brown rot (*Monilinia fructigena*)
	Black spot (*Diplocarpon rosae*)
	Apple scab (*Venturia inaequalis*)
	Ergot (*Claviceps purpurea*)
Imperfect fungi (Deuteromycetes) (over 20 000 species) These fungi lack the ability for sexual reproduction, but are very similar in other respects to ascomycetes	*Penicillium chrysogenum* – an antibiotic (penicillin)
	Dactylaria – traps and feeds on passing nematode worms
	Aspergillus – a mould
	Tichophyton – athlete's foot
	Rhizoctonia – root rots
	Fusarium spp – wilts
Advanced fungi (Basidiomycetes) (over 25 000 species)	Rusts (numerous e.g. *Puccinia graminis*)
	Smuts (numerous e.g. *Ustilago avenae*)
	Bracket fungi
	Mushrooms and toadstools
	Puff balls

Beneficial effects of fungi

Saprophytes: these recycle nutrients. They are able to break down complex organic material and release valuable elements including nitrogen, carbon and sulphur. Fungi are very important in the carbon cycle. Approximately 50 per cent by weight of leaves is carbon. This is released into the atmosphere during decomposition as CO_2 and is therefore available to plants for photosynthesis.

Biotechnology: brewing and baking, steroids, cheese making.

Antibiotics: the production of chemicals which inhibits the growth of competitors, e.g. *Penicillium*.

Weed control: Some fungi attack weeds including wild oats and cleavers. These are called mycoherbicides, e.g. *Colleotrichum* *gloeosporiodes aeschynomene.* New strains are being developed that will destroy drug plants.

Pest control, e.g. *Verticillium lecanii* used to control glasshouse pests.

Control

A fungicide is a chemical control to kill fungi. Soil fungicides are normally applied on nurseries for prophylactic or preventative control, inhibiting the spread of mycelium in the media.

The thermal death point (TDP) is the temperature required to kill an organism. Most fungi have a TDP of 50°C for ten minutes and a few of 80°C for ten minutes. These parameters are considered with reference to soil sterilization procedures.

Life cycles

Figure 20.2 shows a typical life cycle of an advanced fungus. Fungi reproduce from spores released from a fruiting body such as a mushroom. These are less than 0.001 mm and over 100 000 an hour are produced. Few of the spores will germinate successfully and millions are produced to ensure survival. The spores germinate and grow into microscopic cells called hyphae (singular hypha). Groups of hyphae form mycelium strands which continue to grow and spread throughout the soil or into a parasitized plant. Often the vast network of mycelium can extend for several acres in the soil. When environmental conditions are favourable a fruiting body is produced which will release more spores to start the next generation.

Fairy rings case study

Figure 20.3 shows a fairy ring fungus invading turf. Sometimes called 'witches' rings' produced by soil inhabiting fungi *Marasmus* (the

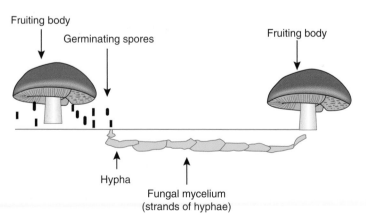

Figure 20.2 Life cycle stages of an advanced fungus

Figure 20.3 Fairy rings

commonest in Britain), *Lycoperdon* or *Hygrophorons* species, the rings of dying grass continue to spread outwards in a concentric pattern up to 30 cm per year for up to two hundred years. The mycelium grows just below the surface and around grass bases, feeding on dead organic matter, forming a dense white fungal growth. Sandy, droughty and poorly fertilized soils seem most susceptible.

At the edge of the ring, lush green grass develops, feeding on released nitrogen as the fungus and microorganisms decompose organic matter. Grass within the ring is often dead or completely bare due to the mass of mycelium growth coating the soil particles. The coating is water-repellent and as a result the soil becomes hydrophobic. Some fairy rings produce hyrogen cyanide and other toxic chemicals which can kill off turf roots. Roots may struggle to acquire water and nutrients in direct competition with the mycelium in the soil, often resulting in drought stress. Turf may recover slowly as the ring gets bigger. Brown toadstools (4–10 cm high) develop between spring and autumn. The mycelium remains dormant over winter.

The fungus is very difficult to control. Raking up toadstools before they release spores can help. The infected turf should be mown and clippings burned separately, disinfecting the lawn mower before proceeding. Old tree stumps and plant roots should be removed before planting new turf. Regular spiking aids water penetration. Watering every two days can restart turf growth within the ring and flushing with sulphate of iron (400 g in 4.5 litres) discourages fungal growth.

An antagonistic form of biological control is now popular. Each ring contains fungus and microorganisms antagonistic to each other. Two fairy rings will not cross each another path because each fungus will produce chemicals which hinder the growth of new mycelium. When the turf is removed and the soil rotivated to 30 cm, the mixed mycelium appear unable to reestablish and newly laid turf establishes well.

Some chemical controls containing *oxycarboxin*, *triforine* and *bupirimate* have been effective at fourteen day intervals. On high value turf areas it may be necessary to excavate and remove the soil to 30 cm depth, refilling with fresh topsoil and reseeding.

Exercise 20.1

Mycelium investigation

Background

Fungi are not made up of cells, but of thread-like hyphae (singular hypha), which contain several nuclei, branching into a network structure. Sometimes cross-walls divide the hyphae. The hyphae spread throughout the soil penetrating between the particles. This network is referred to as mycelium (fungal mycelium). Mycelium are what gives a woodland soil its so called 'earthy' smell. It is believed that fungal mycelium constitutes the largest single living organism in the world, and in the redwood forests of North America have been estimated to extend several kilometers in diameter.

In this exercise fungi specimens (mycelium and spores) have been grown on agar gel, although stale bread, fruit and moist food are equally suitable. Small samples can then be transferred to a microscope slide and examined under a microscope.

Aim

To become familiar with the structure of fungal mycelium.

Apparatus

Monocular microscopes	Needles
White tiles	Bunsen burner
Cotton blue droppers	Slides
Cover slips	Scalpels
Petri dish cultures	

Useful websites

www.fungionline.org.uk

www.tolweb.org/Fungi

www.fungus.org.uk

Method

Select a petri dish containing fungal cultures of *Botrytis*, *Alternaria* and *Penicillium* (or alternative cultured fungi). for each sample complete the following:

1. Take a microscope slide.
2. Use a needle to lift of some of the fungal strands and place on the slide.
3. Place some cotton blue lacto-phenol dye over the slide (sometimes the dye is not necessary).
4. Using two needles carefully separate the strands to make them more clearly visible.
5. Lower a small glass cover slip slowly on top of the droplet using a needle for support.
6. Check that the microscope is on low power.
7. Move the slide around until the fungus is in vision and focus.
8. Move to high power by twisting the lens, looking alongside the microscope to ensure that the lens doesn't break the slide.
9. Focus on the fungus.
10. Draw a labelled diagram of your observations including:
    (a) colour of hyphae
    (b) shape of hyphae
    (c) presence/absence of cross walls
    (d) how spores are held in the hyphae
    (e) shape of spores
    (f) colour of spores.
11. Sterilize your needles in a Bunsen flame between each fungus.

Results

For each fungus draw diagrams of the following:

1. Mycelium strands.
2. Hyphae showing:
    (a) colour of hyphae
    (b) presence/absence of cross walls.
3. Spores showing:
    (a) how spores are held in the hyphae
    (b) colour of spores.

Conclusion

1. When something goes mouldy what is actually happening to it?
2. Explain the terms:
    (a) hyphae
    (b) saprophyte
    (c) parasite
    (d) mycelium.
3. State five diseases causes by fungus.
4. Draw a labelled diagram to show a small portion of a fungal hyphae.
5. In what ways to fungi affect the natural world?

Exercise 20.2

Spore cases and germinating spores

Background

Fungi reproduce not by seed but from spores. The spores are supported on a special extended stalk tip of vertical hyphae called a sporangia. The spores are encased in a sporangium and are released into air currents when its wall breaks down. The ascomycete and basidomycetes fungi group reproduce sexually and their spore case is called a perithecia, but otherwise fulfils the same function as a sporangium. Once the spores have been transported to a new sight such as a leaf blade, they germinate and grow into the host material, forming new mycelium fibres.

Fungi are very fast growers. For example, it has been estimated that in twenty-four hours a half tonne cow will manufacture 0.5 kg of protein, but half a tonne of yeast will grow to 50 tonnes and only uses up a few square metres of soil. For this exercise spores may be collected from any suitable fungus and prepared on a microscope slide.

Aim

To investigate the method of fungus reproduction.

Apparatus

Microscopes
Slides of germinating spores

Useful websites

www.biotopics.co.uk/microbes/fungi.html

www.fungionline.org.uk

Method

1. Place the slide under the microscope and focus.
2. Squash the spore case gently with a needle to encourage spore emergence.
3. Draw a labelled diagram of your observations.

Results

Draw a diagram of emerging/germinating spores.

Conclusions

1. How large are the spores?
2. How many spores are produced?
3. Name two types of spore casing produced by fungi.

Exercise 20.3

Spore cases and spore survival techniques

Background

Some spores are able to survive over winter until conditions for germination are improved. This normally involves having a harder resistant casing (sclerotia) as in botrytis. Using this knowledge horticulturalists are able to apply preventative fungicides prior to spore germination.

Aim

To investigate spore survival techniques.

Apparatus

> *Botrytis sclerotia*
> Microscope slides
> Microscopes

Useful websites

www.fungionline.org.uk

www.rhs.org.uk/advice/profiles0406/botrytis.asp

http://plantclinic.cornell.edu/FactSheets/botrytis/botrytis_blight.htm

Method

1. Observe the sclerotia (spore case) of botrytis.
2. Draw a labelled diagram of your observations.
3. Try to break open the case.

Results

Draw a diagram of your observations.

Conclusion

1. Suggest two advantages the fungus would derive from producing sclerotia.
2. Explain why fungi are 'cryptogamous' organisms.
3. What role do fungi play in the decomposition of materials?
4. Explain the difference between a 'facultative' and 'obligate' parasite.

Exercise 20.4

Rust disease

Background

Several diseases can result from fungi including blight, mildew and rust.

There have been several fungal epidemics including the Irish potato famine of the 1840s which was caused by the potato blight fungi (*Phytophthora parasitica*) favoured by high temperatures and humidity. Dutch elm disease is also caused by a fungus carried by beetles from tree to tree.

Rust is a common and easily identified fungal disease. Susceptible plants include, particularly, groundsel, plums, beans, carnations, pelargoniums and chrysanthemums. The mycelium causes the leaf to have a ragged appearance around a raised dark brown leaf spot containing spores that can be brushed off with your fingers.

Aim

Recognition of rusts.

Apparatus

Rust-infected plant
Microscope

Useful websites

www.biology.ed.ac.uk/research/groups/jdeacon/FungalBiology/rust.htm

www.fungionline.org.uk

Method

1. Observe the specimen under the microscope.
2. Draw a labelled diagram of your observations.

Results

Draw a labelled diagram of your observations.

Conclusions

1. Which group of fungi do rust belong to?
2. List some plants susceptible to rust disease
3. Which part of the plant does rust fungus normally attack?
4. Name four fungal disease from the ascomycetes group.

Exercise 20.5

Powdery mildew disease

Background

These intermediate fungi infect leaves. Susceptible plants include, particularly, cucumber, roses and apples, although many country hedgerows also appear to be good hosts. The fungus sits on the upper surface of the leaf and inserts short feeding probes through which food is sucked up. They favour hot, dry conditions and are particularly important because unlike most other fungi they do not require a wet surface to begin infecting.

Aim

Recognition of mildew.

Apparatus

Mildew-infected plant
Microscope

Useful websites

www.hgca.com/hgca/wde/diseases/Mildew/Milcyc.html

www.rhs.org.uk/advice/profiles0800/powdery_mildews.asp

www2.defra.gov.uk

www.fungionline.org.uk

Method

1. Observe the specimen under the microscope.
2. Draw a labelled diagram of your observations.

Results

Draw a labelled diagram of your observations.

Conclusion

1. Which group of fungi do mildew belong to?
2. List some susceptible plants.
3. What conditions stimulate powdery mildew growth?
4. How is powdery mildew different from most other fungi?
5. Name four fungal diseases from the primitive fungi group.

Exercise 20.6

Structure of a mushroom

Background

The advanced fungi reproduce by spores that are broadcast from a specialized hyphae called a 'fruiting body' such as a mushroom. Many people think that a mushroom is an individual plant, but in reality it is only part of a much more extensive mycelium organism living in the soil. Bracket fungi, such as chicken-in-the-wood, which grow outwards from trees reveal to trained horticulturalist that the whole tree will be infected by mycelium and will be slowly dying. When ripe, the gills turn brown and the spores are carried away in the wind.

Aim

To investigate the fruiting body structure.

Apparatus

Microscope
Parasol mushroom

Useful websites

www.fungionline.org.uk

www.world-of-fungi.org/Mostly_Mycology/Richard_Clarke/intro.htm

www.geocities.com/rainforest/andes/8046/structure.html

Method

1. Observe the fruiting body.
2. Draw a labelled diagram.
3. Remove the stalk.
4. Observe the gills under the microscope and report any significant observations.

Results

Draw a diagram of the fruiting body and note your observations.

Conclusions

1. Suggest why toadstools may be found growing in dark areas of woodland where green plants cannot survive.
2. Explain what is meant by the term 'fruiting body'.
3. Which group of fungi are responsible for the production of 'fruiting bodies'?
4. State two environmental conditions that would favour the establishment of a fungus spore on a leaf surface.
5. Name two fungal disease, the plants they attack and the symptoms they produce.

Exercise 20.7

Spore mapping

Background

The field mushroom (*Agaricus campestris*) is estimated to produce 100 000 spores per hour. Each spore is only 0.001 mm in diameter.

The mushroom itself is only the fruiting body of a much larger organism forming the mycelium living underground. There are over 80 000 species of fungus and identification (mycology) is a major science in it self. Often many species can only be separated by reference to a spore map. This is produced by placing a fungus fruiting body, gills downward, on a piece of white card and observing the resulting spore pattern and colouring that develops as the gills pump out spores.

Aim

To produce a spore map of a fungus.

Apparatus

Fungus fruiting body
White paper

Useful websites

www.fungionline.org.uk

www.nifg.org.uk/recording_tips.htm

Method

1. Label the top right-hand corner of the paper with specimen details.
2. Remove the stalk from a mushroom with dark brown gills (i.e. mature).
3. Place the cap on a piece of paper, gills facing downwards.

4. After forty-eight hours discard the fungus; a spore map will have been produced.
5. Collect your map for reference.

Results

Paste in spore map in a notebook. This may be preserved for future reference by covering with transparent plastic.

Conclusions

1. What can you conclude about the rate of production of spores?
2. In what way would your map have differed if left for longer?
3. What advantage does a fungus yield from producing an abundance of spores?
4. State four positive benefits of fungus to horticulturalists.

Answers

Exercise 20.1. Mycelium investigation

Results

Simple diagrams should be presented showing spores, hyphae and mycelium indicating the presence or absence of cross walls.

Conclusions

1.   The presence of mould indicates that organic matter is being parasitized by fungi as part of the nutrient recycling process. Ultimately, the entire body will be digested and disintegrate into nothing.

2.   (a) Hyphae are individual cells of fungi.

(b) Saprophytes feed on dead and decaying material.

(c) Parasites derive their nutrition from living material.

(d) Mycelium are threads of hyphae joined together.

3.   Any five fungal diseases, e.g. *Phytophthora*, *Pythium*, *Rhizoctonia*, *Fusarium* and *Botrytis*.

4.   Appropriate diagram as in Figure 20.1.

5.   Fungi are the decomposers of the natural world. They decompose organic matter and recycle nutrients.

Exercise 20.2. Spore cases and germinating spores

Results

Appropriate diagram.

Conclusions

1.   Spores are typically 0.001 mm in diameter.

2.   The number on the slide may be counted. At least eight spores per case should be seen.

3.   Sporangia and perithecia.

Exercise 20.3. Spore cases and spore survival techniques

Results

Appropriate diagram.

Conclusions

1.   An ability to survive over winter and for several years in the soil.

2.   Fungi reproduce by spores and are therefore crytogamous organisms.

3.   Fungi are the primary decomposers in the natural world, digesting organic matter and recycling nutrients.

4.   Facultative parasites prefer to derive their nutrition from dead matter but can feed on living material. Obligate parasites feed on only dead matter.

Exercise 20.4. Rust disease

Results

Appropriate diagram.

Conclusions

1.   Advanced fungi (Basiodomycetes).

2.   Groundsel, plums, beans, carnations, pelargoniums and chrysanhemums.

3.   Leaves.

4.   Four examples, e.g. powdery mildew, leaf spots, Dutch elm disease, and black spot.

Exercise 20.5. Powdery mildew disease

Results

Appropriate diagram.

Conclusions

1.   Intermediate fungi (ascomycetes).

2.   Cucumber, roses and apples.

3.   Hot, dry conditions.

4.   Powdery mildew does not require wet conditions for spores to germinate.

5.   Four examples e.g. club root, damping-off, downy mildew and potato blight.

Exercise 20.6. Structure of a mushroom

Results

Appropriate diagram.

Conclusions

1.   Toadstools and other fungi do not contain chlorophyll and do not require light for growth. Instead they derive their nutrition from decomposing organic matter.

2.   Fruit body is the specialist spore-producing glands, often appearing as a mushroom structure.

3.   Only advanced fungi produce fruiting bodies.

4.   Wet, damp conditions, high humidity and still air.

5.   Suitable diseases, e.g. damping-off affecting seedlings and rust affecting chrysanthemums.

Exercise 20.7. Spore mapping

Results

A suitable spore map recorded.

Conclusions

1.   A mass of spores are produced, typically over 100 000 per hour.

2.   More spores would have produced a more detailed map.

3.   Increases likelihood of species survival.

4.   Four appropriate examples, e.g. saprophytes, biotechnology, antibiotics, mycoherbicides and pest control.

Chapter 21 Insects and mites

Key facts

1. Most pests come from the order Arthropda, of which insects are the most damaging.
2. Insect pests may be either winged or wingless.
3. In diagnosing a pest, first decide whether the damage is by biting/chewing or sap-sucking, then focus in on more specific symptoms of pest attack.
4. Only by understanding and interrupting the pest life cycle can we gain control.
5. Good intelligence such as from daily observations, pit fall trapping and pooter bug collecting, is a vital key in pest management strategies.
6. Pests may be controlled by cultural, biological and chemical methods.
7. When using chemical controls always check approval status through the current edition of *The UK Pesticide Guide*.

Background

Insects and mites can be pests of horticultural crops, normally by feeding on the plant as a source of nutrition. The animal kingdom is enormous and it would be impossible to address every single pest. The exercises in this chapter are designed to provide an introduction to pest recognition and management through investigating pest life cycles, together with the damage they do to plants and some chemical controls.

Typical pests of horticultural plants, which come mostly from the order arthropoda, contributing 80 per cent of all known organisms, include some crustaceans (e.g. woodlice), arachnida (e.g. scorpions, mites, daddy long legs), diplopoda (millipedes), chilpoda (centipedes) and insecta (insects).

Of the arthropods, the insects are the most important. There are three million known invertebrate insect species. Their general characteristics are that they have a hard exoskeleton or cuticle made from chitin and proteins, segmented bodies and jointed legs. Insects differ from other arthropods because they have three pairs of legs and a body divided into a head, thorax, and abdomen. They may be simply divided into winged (pterygota) or wingless (apterygota) insects. Wingless insects include pests such as silverfish, springtails and bristletails. The winged insects are easily the largest group of pests and include organisms such as most bugs, beetles, butterflies, moths, ants, wasps, lace-wings and thrips.

Figure 21.1 Symptoms of Thrip damage

Generally, the damage that these pests do to plants can be grouped into whether it is by gnawing, biting or sap-sucking. For example, the thrip (*Thysanoptera*), otherwise known as a thunder bug or gnat, has a long feeding probe which it inserts into the petals of flowers and feeds on the sap. As it does so, it cause air to rush into the cell, creating a characteristic silver-white streaking on the leaves and petals. For further symptoms of pest damage see Figure 21.1.

Specimen collection

Often, it is a particular phase in the life cycle of a pest which causes damage to plants. It will be useful to collect all life cycle stages, including grubs and adults (see Figure 21.2). For example, it is the larvae stage (leatherjackets) of a daddy long legs (cranefly), that feeds close to the surface of the soil, damaging roots and underground parts of the stem of grasses.

In certain cases it may be useful to collect several pupae (chrysalides, singular chrysalis) and allow one to hatch, so that a complete life cycle specimen collection can be built up.

Identification will also be aided by reference to a simple key. Organisms are often characterized by their number of legs and other easily observed features.

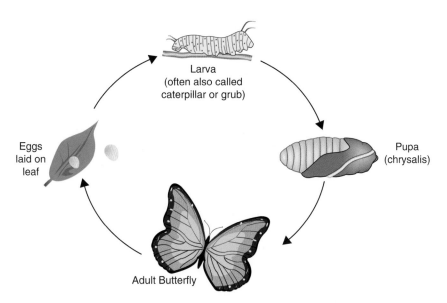

Figure 21.2 Typical pest life cycle stages

In this chapter the following pests will be investigated: vine weevils, wireworms and click beetles, leatherjackets and cranefly, cabbage root fly, yellow underwing moth, common gooseberry sawfly, cabbage white butterfly, red spider mite and aphids. In addition chemical insecticide controls will be stated through reference to *The UK Pesticide Guide* (the 'green book'; Whitehead, 2008), previously introduced in Chapter 4.

Exercise 21.1

Pit fall trapping

Background

Despite their abundance it can be a difficult task to collect pests for investigation. Two methods are presented to enable this to be done. First, the creation of a pit fall trap. Second, bug hunting with a collecting jar called a pooter (see Exercise 21.2). In addition, rummaging through compost and old heaps often yields plenty of 'beasties' for study.

The use of a pit fall trap is especially convenient since it requires only a little work to make and install. It involves inserting a plastic container into the ground into which beasties, in their inquisitiveness, fall. A clean half pint plastic beer glass has been found to be ideal for this purpose. The insects are drowned in a little water in the base of the container and collected weekly for identification. This can be continued on a permanent basis throughout the year so that different life cycles are collected; it and also helps, to some degree, in assessing pest numbers.

Aim

To enable common pests to be pit fall trapped for study and private collections.

Apparatus

Plastic container
A few pebbles
Water
Wire mesh

Useful websites

www.bna-naturalists.org/kids/trees.htm

www.fieldworklib.org/asp/print.asp?type=k&cat=1&ks=3&id=168&secID=452

www.thenakedscientists.
com/HTML/content/kitchenscience/exp/catching-insects-with-pitfall-traps

Method

Set up the apparatus as shown in Figure 21.3.

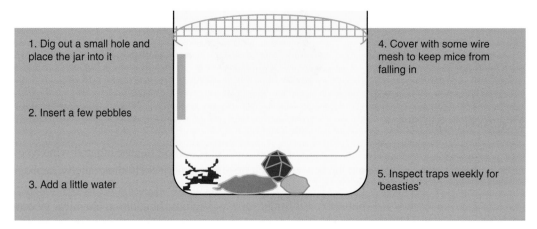

1. Dig out a small hole and place the jar into it

2. Insert a few pebbles

3. Add a little water

4. Cover with some wire mesh to keep mice from falling in

5. Inspect traps weekly for 'beasties'

Figure 21.3 Typical pit fall trap

Results

Record the type and number of organisms found by reference to a simple key.

Conclusions

1. What sort of organisms are pests in horticulture?
2. Describe typical insect feeding methods that may damage plants.
3. Explain the difference between a 'larva' and a 'pupa'.

Exercise 21.2

Pooter bug hunting

Background

A pooter is a devise used to collect organisms without needing to touch them directly (see Figure 21.4). Many insects are difficult to catch and a pooter is

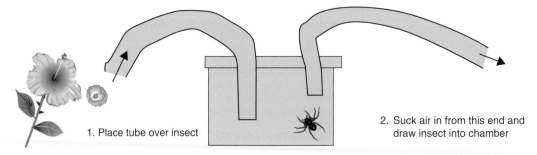

1. Place tube over insect

2. Suck air in from this end and draw insect into chamber

Figure 21.4 Typical home-made pooter

often useful to trap them before they have a chance to escape. Old compost and the leaves of most trees and shrubs are a good source of insects.

Aim

To investigate and collect pests found feeding using a hand-held pooter.

Apparatus

Pooter

Useful websites

www.show.me.uk/site/make/Natural-World/ACT59.html

http://bugguide.net

Method

1. Rummage through old compost or plant material and collect any bugs found using a pooter.
2. Identify the insects using a suitable key.

Results

Identify and record the type (by reference to a key), and number of organisms found.

Conclusions

1. What sort of insects were mostly found feeding on old compost?
2. What sort of insects were mostly found feeding on leaves?
3. Explain what is meant by the term 'life cycle'.
4. State the life cycle stages of the 'beasties' that you trapped.

Exercise 21.3

Vine weevils (*Otiorhynchus*)

Background

The vine weevil (*Otiorhynchus sulcatus*) is one of the common species of weevil and is a regular plague suffered by many households, gardens and nurseries. Their first pair of wings are fused together for protection, largely for an underground lifestyle. As a result they are flightless, walk rather slowly and are often referred to as wingless weevils. They damage plants by biting, chewing and gnawing at roots, stems, leaves and buds. Almost all plants are susceptible but especially, *Azalea, Begonia, Camellia, Cissus, Cyclamen, Gloxinia, Hydrangea, Impatiens, Kalanchoe, Pelargonium, Primula, Rhododendron, Skimmia, Taxus, Thuja, Viburnum,* together with alpines, conifers, roses, tree fruit, fruit bushes and strawberries. The damage caused is often confused with diseases especially damping-off and *Rhizoctonia* root rots.

Life cycle

The adult sometimes attacks leaves, but the larva stage is the most damaging. Always inspect roots for signs of the grub (1 cm long, slightly curved, legless white body with a chestnut brown head) (see Figure 21.5).

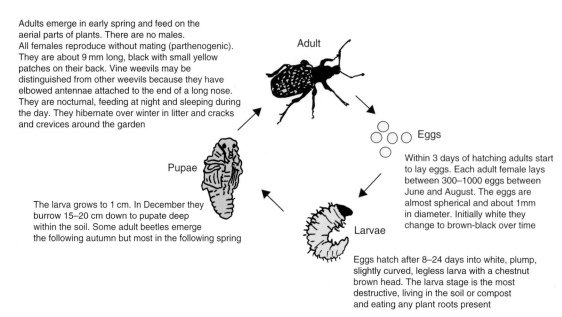

Adults emerge in early spring and feed on the aerial parts of plants. There are no males. All females reproduce without mating (parthenogenic). They are about 9 mm long, black with small yellow patches on their back. Vine weevils may be distinguished from other weevils because they have elbowed antennae attached to the end of a long nose. They are nocturnal, feeding at night and sleeping during the day. They hibernate over winter in litter and cracks and crevices around the garden

Adult

Pupae

Eggs

Within 3 days of hatching adults start to lay eggs. Each adult female lays between 300–1000 eggs between June and August. The eggs are almost spherical and about 1mm in diameter. Initially white they change to brown-black over time

The larva grows to 1 cm. In December they burrow 15–20 cm down to pupate deep within the soil. Some adult beetles emerge the following autumn but most in the following spring

Larvae

Eggs hatch after 8–24 days into white, plump, slightly curved, legless larva with a chestnut brown head. The larva stage is the most destructive, living in the soil or compost and eating any plant roots present

Figure 21.5 Life cycle of a vine weevil

Visual symptoms

Underground damage signs include barked root stock, holes in corms and tubers and plants that are easily lifted from the ground due to severed roots.

The leaves turn yellow and have characteristic half-moon shapes notched out by the feeding adult. Stems show flagging growth and wilting. Flower buds may be destroyed and feeding causes premature senescence.

Control methods

Today, horticulturalists no longer aim for pest eradication, but rather towards pest management, keeping numbers below levels which cause significant damage.

Pests may be controlled by either cultural, biological, chemical or integrated pest management strategies.

Cultural control

Techniques of good husbandry. Early detection is the basis of control, followed by manual removal, repotting of plants into fresh compost and good hygiene practice.

Regular pest monitoring is important. Look for adults sleeping under pot lips and hunt adults at night. Use pit fall traps and rolled-up pieces of sacking or corrugated paper near infected plants. These should be inspected every few days and any weevils destroyed. Thoroughly clear the garden before replanting and cultivate the soil to kill any grubs. Examine new plants thoroughly for signs of attack and plant in late August after the main egg-laying period. Any infected plants should be discarded.

Biological control

Two specific biological control pathogens are known for vine weevil: *Metarrhizum anisopliae,* an entopathogenic fungus which infects young eggs; and parasitic nematodes (*Heterorhabditis megidis* and *Steinernema carpocapsae*)

which penetrate the larva through mouth, anus or respiratory openings. They release bacteria as they feed. As the bacteria multiply the larva turn brownish-red and die from poisoning. However, to work effectively the soil temperature should be above 14°C. Example products include *Grubsure, Nature's Friends and Fightagrub.*

Chemical control

Vine weevils are resistant to many insecticides.

Chemicals available include the following active ingredients: fipronil, a systemic insecticide for soil treatment; chloropyrifos, a contact and ingested insecticide; and imidacloprid, an insecticide for soil and foliar treatment. As vine weevils are nocturnal, the best time to spray a contact insecticide is at night after the adults have climbed up on to the leaves. Death follows 2–3 days after spraying.

Levington's Plant Protection Compost and *Container & Hanging Basket Compost with Vine Weevil Control* is very popular to control the weevil. It is a compost ready mixed with a systemic pesticide called *Intercept* which contains the active ingredient imidacloprid. As plants grow they absorb *Intercept* into their tissue. It is toxic to vine weevil but harmless to plants. Any vine weevil that subsequently feeds on plants grown in the compost are poisoned and die. The plants grow on happily. It also kills larvae and adults in the root zone on contact with the chemical.

The compost is safe to handle. Intercept is also biodegradable and harmless to beneficial insects.

Integrated pest management (IPM)

All these techniques (cultural, biological and chemical) may be combined into an integrated pest management (IPM) programme aimed at keeping pest numbers at a level where plant damage is minimal.

Aim

Investigation of common horticultural pests.

Apparatus

Larvae, pupae and adult vine weevil.
The UK Pesticide Guide (Whitehead, 2008).

Useful websites

www.rhs.org.uk/advice/profiles0600/vineweevil.asp

www.cyclamen.org/weevil_set.html

www.defenders.co.uk/pest-problems/vine-weevil.html

www.pesticides.gov.uk

www.ukpesticideguide.co.uk

Method

1. Observe the specimens.
2. Draw the life cycle stages.

Results

1. State what damage to plants each of the life cycle stages causes and whether by biting or sucking:
   (a) larvae
   (b) pupae
   (c) adult.

Conclusions

Using *The UK Pesticide Guide*, state what chemicals can be used to control the pest. Give the following details:
(a) active ingredients
(b) product names
(c) manufacturing company.

Wireworms and click beetles (*Agriotes, Athous and Ctenicera*)

Background

Wireworms are probably the best-known soil-inhabiting plant pests. Click beetles are named from the sound (click) they make as they flick themselves into the air to right themselves after falling on their back. About sixty species are found in Britain, the larvae of which are called wireworms.

Susceptible plants include grass, cereals, potatoes, sugar beet, brassicas, dwarf French beans, lettuce, onions, strawberries, tomatoes and anemones,

Aim

Investigation of common horticultural pests.

Apparatus

larvae, and adult
The UK Pesticide Guide (Whitehead, 2008)

Useful websites

www.dgsgardening.btinternet.co.uk

www.ukpesticideguide.co.uk

Method

1. Observe the specimens.
2. Draw the life cycle stages.

Results

1. State what damage to plants each of the life cycle stages causes, and whether by biting or sucking
   (a) larvae
   (b) pupae
   (c) adult.
2. When are the adult beetles formed?
3. When are the eggs laid in the soil?
4. How many pairs of legs have wireworms got?

Conclusion

Using *The UK Pesticide Guide*, state what chemicals can be used to control the pest. Give the following details:

(a) active ingredients
(b) product names
(c) manufacturing company.

Leatherjackets and cranefly (*Diptera, Tipula*)

Background

The sworn enemies of turf managers, cranefly (also called daddy long legs) larvae are known as leatherjackets and thrive especially in prolonged damp weather in late summer. Their presence is often evident from a mass of birds, especially rooks, feeding on the grass. Leatherjackets hatch in autumn and immediately begin feeding on turf roots. The following summer they pupate, and adult cranefly emerge.

Susceptible plants include grass, clover, strawberries, cereals, sugar beet, brassicas, courgettes and herbaceous garden plants. Parasitic nematodes have been shown to give some measure of control over this pest.

Aim

Investigation of common horticultural pests.

Apparatus

    All life cycle stages
    The UK Pesticide Guide (Whitehead, 2008).

Useful websites

www.rhs.org.uk/advice/profiles0206/leatherjackets.asp

www.ukpesticideguide.co.uk

Method

1. Observe the specimens.
2. Draw a labelled diagram of the cranefly structure.

Results

1. State what damage to plants each of the life cycle stages causes and whether by biting or sucking:
   (a) larvae
   (b) pupae
   (c) adult.
2. What are cranefly also known as?
3. How many legs do leatherjackets have?

Conclusions

Using *The UK Pesticide Guide*, state what chemicals can be used to control the pest. Give the following details:

(a) active ingredients
(b) product names
(c) manufacturing company.

Exercise 21.6

Cabbage root fly (*Delia radicum*)

Background

A destructive pest of Brassicaceae (previously Cruciferae family) plants found in the vegetable garden, the white, 1 cm long larvae attack seedlings and mature plants, damaging the root system. Susceptible plants are cauliflowers, broccoli, cabbage, Brussels sprouts, calabrese, Chinese cabbage, radish, swede, turnip, garden stocks and wallflowers. Symptoms include stunted growth, seedling die-off, discoloured blue and wilting leaves, black and rotten roots.

Aim

Investigation of common horticultural pests.

Apparatus

life cycle stages: larvae, pupae and adult
The UK Pesticide Guide (Whitehead, 2008)

Useful websites

www.abdn.ac.uk/organic/organic_21.php

www.ukpesticideguide.co.uk

Method

1. Observe the specimens.
2. Draw a labelled diagram of the life cycle.

Results

1. State what damage to plants each of the life cycle stages causes and whether by biting or sucking:
   (a) larvae
   (b) pupae
   (c) adult.
2. What colour is the adult fly?
3. How long is the fly?
4. When can the first eggs be found in plants?
5. What colour are the maggots?
6. How many legs have the maggots?

Conclusions

Using *The UK Pesticide Guide*, state what chemicals can be used to control the pest. Give the following details:

(a) active ingredients
(b) product names
(c) manufacturing company.

Exercise 21.7

Cutworms

Background

Cutworm is the name given to caterpillars of moths which 'cut-down' plant stems at ground level, often severing them from the tap root. Moths which produce caterpillars known as 'cutworms' include the turnip moth (*Agrotis segetum*), garden dart moth (*Euxoa nigricans*), Whiteline dart moth (*Euxoa tririci*) and the yellow underwing moth (*Noctua pronuba*).

Susceptible plants include lettuce, beet, beetroot, celery, carrots leeks, potatoes, turnips, swedes, strawberries and other plants with tender tap roots.

Aim

Investigation of common horticultural pests.

Apparatus

Life cycle stages
The UK Pesticide Guide (Whitehead, 2008)

Useful website

www.ukpesticideguide.co.uk

Method

1. Observe the specimens.
2. Draw a labelled diagram of the life cycle.

Results

1. State what damage to plants each of the life cycle stages causes and whether by biting or sucking:
   (a) larvae
   (b) pupae
   (c) adult
2. How long are the larvae?
3. What colour are the larvae?
4. How many legs do the larvae have?

Conclusions

Using *The UK Pesticide Guide*, state what chemicals can be used to control the pest. Give the following details:

(a) active ingredients
(b) product names
(c) manufacturing company.

Exercise 21.8

Common gooseberry sawfly (*Nematus ribesii*)

Background

The caterpillars of this sawfly feed on the leaves of red and white currants and gooseberries. The rose-leaf rolling sawfly (*Blennocampa pusilla*) attacks leaves of rose plants. Larvae feed on both leaves and fruit. They are easy to identify from their light green bodies covered in black spots.

Aim

Investigation of common horticultural pests.

Apparatus

Life cycle stages
The UK Pesticide Guide (Whitehead, 2008)

Useful websites

www.rhs.org.uk/advice/profiles0601/gooseberry_sawfly.asp

www.rhs.org.uk/advice/profiles0905/rose_sawfly.asp

www.ukpesticideguide.co.uk

Method

1. Observe the specimens.
2. Draw a labelled diagram of the life cycle.

Results

1. State what damage to plants each of the life cycle stages causes and whether by biting or sucking:
   (a) larvae
   (b) pupae
   (c) adult.
2. How many pairs of wings do the adults have?
3. What colour are the larvae?
4. How long are the larvae?

Conclusions

Using *The UK Pesticide Guide*, state what chemicals can be used to control the pest. Give the following details:

(a) active ingredients
(b) product names
(c) manufacturing company.

Exercise 21.9

Cabbage white butterfly (*Mamestra brassicae*)

Background

The white butterfly caterpillars feed on the aerial parts of almost all Brassicaceae (previously Cruciferae) plants, garden nasturtiums and general allotment plants, causing severe defoliation. There are two types: the large white (*Pieris brassicae*) and the small white (*Pieris rapae*). Adults lay eggs on the leaves. As soon as the larvae hatch they begin feeding on the leaves and tunnelling into vegetables. The vegetable gardener needs to pay daily attention to this pest, removing both eggs and caterpillars. Ensure plants are covered with netting and encourage feeding birds into the garden.

Aim

Investigation of common horticultural pests.

Apparatus

Life cycle specimens
The UK Pesticide Guide (Whitehead, 2008)

Useful websites

www.rhs.org.uk/advice/profiles0800/cabbage_caterpillars.asp

www.dgsgardening.btinternet.co.uk/cabwhitelge.htm

www.ukpesticideguide.co.uk

Method

1. Observe the specimens.
2. Draw a labelled diagram of the life cycle.

Results

1. State what damage to plants each of the life cycle stages causes and whether by biting or sucking:
   (a) larvae
   (b) pupae
   (c) adult.
2. When do the females lay their eggs?
3. How long are the larvae?
4. What colour are the larvae?
5. In which month do the larvae start to pupate?

Conclusions

Using *The UK Pesticide Guide*, state what chemicals can be used to control the pest. Give the following details:

(a) active ingredients
(b) product names
(c) manufacturing company.

Exercise 21.10

Red spider mite (*Tetranychus urticae*)

Background

Also known as the two-spotted spider mite and rarely looking red, is a pest on protected crops but can damage outdoor plants in hot, dry summers. The mites feed on the underside of leaves, sucking the sap and also spinning a fine silken web over the plant. The females turn red after mating in the autumn. They may be controlled biologically using another mite called *Phytoseiulus persimilis*, which is, in fact, a lot redder than red spider mite itself (see Chapter 23). The mites are too small to be seen with the naked eye. It is their fine web and leaf speckling that exposes their presence.

Susceptible plants include all glasshouse crops, strawberries, hops, beans, black currants, cane fruit, hops, fruit trees and ornamentals.

Aim

Investigation of common horticultural pests.

Apparatus

Infected plants
Binocular microscope/hand lens
The UK Pesticide Guide (Whitehead, 2008)

Useful websites

www.rhs.org.uk/advice/profiles0601/red_spider_mite.asp

www.ukpesticideguide.co.uk

Method

1. Observe the specimens under the microscope.
2. Draw a simple diagram of the mite's life cycle.

Results

1. How many legs have the mites?
2. How big are they?
3. State the damage to plants caused by the adult life cycle stage and whether by biting or sucking:
    (a) larvae
    (b) pupae
    (c) adult.

Conclusions

1. Using *The UK Pesticide Guide*, state what chemicals can be used to control the pest. Give the following details:
    (a) active ingredients
    (b) product names
    (c) manufacturing company.
2. State the predator often used in biological control of this pest.
3. Explain what is thought to cause the females to turn red.

Exercise 21.11

Aphids (*Hemiptera*)

Background

Aphids are an example of a sap-sucking pest. They are soft-bodied, pear-shaped bugs from the order *Hemiptera*. Often called greenfly, blackfly or plant louse, they vary considerably in size and colour from green, black, grey to orange. The green-coloured aphid only appears in midsummer. There are several hundred species in Britain, including peach-potato aphid (*Myzus persicae*) and melon-cotton aphid (*Aphis gossypii*).

Figure 21.7 shows the typical life cycle of an aphid. They damage plants by piercing leaf and petal tissues and feeding on the sap. They do not suck the sap; it is forced into the feeding stylet by cell pressure. Consequently they absorb more sugar than they need and the excess is secreted as honey dew from tubes on their back. The honey dew landing on leaves is often colonized by sooty mould and other fungal diseases. Aphids feed on all salad, flower crops, bedding plants, pot plants, herbaceous plants, vegetables, trees and shrubs.

Figure 21.6 Aphids

The visual symptoms of aphid damage include curled and distorted growth, poor quality flowers, sticky honey dew, reduced photosynthesis, ants feeding on honey dew, sooty mould fungus and viruses (over 100) passed on during feeding.

Control methods

Cultural control

Early detection is the basis of control and regular monitoring using yellow sticky traps is helpful. All cuttings and plant material should be inspected

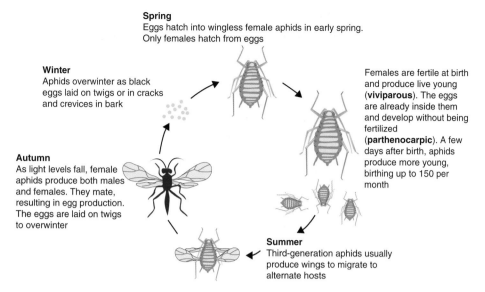

Spring
Eggs hatch into wingless female aphids in early spring. Only females hatch from eggs

Winter
Aphids overwinter as black eggs laid on twigs or in cracks and crevices in bark

Females are fertile at birth and produce live young (**viviparous**). The eggs are already inside them and develop without being fertilized (**parthenocarpic**). A few days after birth, aphids produce more young, birthing up to 150 per month

Autumn
As light levels fall, female aphids produce both males and females. They mate, resulting in egg production. The eggs are laid on twigs to overwinter

Summer
Third-generation aphids usually produce wings to migrate to alternate hosts

Figure 21.7 Life cycle of an aphid

Figure 21.8 Ladybirds used to control aphids

on purchase, together with removing weeds which can act as an alternate host for aphids. Plants should be well watered in dry weather so that cells are turgid. Common practice has also been to collect ladybirds and place them on plants under attack. Success has also been found from hanging fat from garden shrubs. Birds that are waiting to feed, will hunt for aphids, as well as planting flora to lure beneficial natural enemies to the garden.

Chemical control

Aphids have developed much resistance to many modern synthetic pesticides. Some insecticides used are pyrethrins, pirimicarb, cypermethrin, nicotine, aldicarb, and soap solution. Avoid chemicals which may harm biological control agents.

Biological control

Choice of parasite, predator or parasite are available (see Chapter 23). Natural enemies are very specific about which species of aphid they feed on. In nature the explosive increase in aphids is checked by parasitic fungi and by predators such as ladybirds, lacewings, hoverflies, syrphid flies, green lacewings, lady beetles, birds, spiders and gall midges. Some commercially available products include: *Delphastus pusillus* (predatory beetle), *Aphidius matricariae* (parasitic wasp), *Hipodamia convergens* (predatory ladybird), *Chrysoperla carnea* (predatory lacewing larvae) and *Verticillium lecanii* (a pathogenic fungal disease).

Aim

Investigation of common horticultural pests.

Apparatus

Aphid-infected plants
Binocular microscope/handlens
The UK Pesticide Guide (Whitehead, 2008)

Useful websites

www.rhs.org.uk/advice/profiles0406/aphids.asp

www.ukpesticideguide.co.uk

Method

1. Observe the specimens under the microscope
2. Draw a simple diagram of the insect's life cycle.

Results

1. How many legs have the aphids?
2. How big are they?
3. State what damage to plants each of the life cycle stages causes and whether by biting or sucking:
   (a) larvae
   (b) pupae
   (c) adult.

Conclusions

Using *The UK Pesticide Guide*, state what chemicals can be used to control the pest. Give the following details:

(a) active ingredients
(b) product names
(c) manufacturing company.

Answers

Exercise 21.1. Pit fall trapping

Results

A range of soil organisms will be found similar to those presented in Figure 16.1.

Conclusions

1. Any organism that damages plant structure such as beetles, caterpillars and mites.

2. Damage will either be by biting/chewing or by sap-sucking insects that insert a stylet into plant tissue.

3. A larva is the juvenile life cycle stage of an insect, sometimes called a caterpillar or grub. A pupa is the chrysalis stage encasing the insect going through metamorphosis from which the adult will emerge.

Exercise 21.2. Pooter bug hunting

Results

A range of soil organisms similar to those in Figure 16.1 will be found.

Conclusions

1. Soil-inhabiting organisms, beetles, mites and other biting/chewing insects.

2. Caterpillars, whitefly, red spider mites and other sap-sucking insects.

3. Life cycle refers to the stages in the development of an organism from birth to death.

4. Typically for a beetle: egg → larva→ pupa →adult.

Exercise 21.3. Vine weevils

Results

1. (a) Major damage by biting/chewing mouth parts feeding on mainly on plant roots.

    (b) None.

    (c) Adult. Feed on aerial parts eating leaves and leaving a U-shaped notch. Biting mouth parts.

2. The snout is particularly long with elbowed antennae.

3. 9 mm.

4. Within 3 days of hatching from June to August.

5. In winter months.

Conclusions

For example:

(a) imidacloprid

(b) Intercept Levington's plant protection compost

(c) Scotts company.

Exercise 21.4. Wireworms and click beetles

Results

1. (a) biting and chewing damage to underground plant parts

    (b) none

    (c) little damage.

2. September, then they hibernate.

3. May through July.

4. Three pairs.

Conclusions

For example:

(a) ethoprophos

(b) Mocap 10 Gl

(c) Bayer Crop Science.

Exercise 21.5. Leatherjackets and cranefly

Results

1. (a) Leatherjackets feed close to soil surface, biting and chewing underground stems and roots.

    (b) None.

    (c) Daddy long legs, but adult are harmless to plants.

2. Daddy long legs.

3. None

Conclusions

For example:

(a)   chlorpyrifos

(b)   Suscon Green soil insecticide

(c)   Fargo Ltd.

Exercise 21.6. Cabbage root fly (*Delia radicum*)

Results

1. (a) Damage plant roots by biting/chewing between April and May

    (b) None

    (c) None.

2. Dark grey

3. 6 mm

4. April/May/

5. White or cream coloured

6. None, fly larvae don't have legs

Conclusion

For example:

(a)   carbosulfan

(b)   Marshall Soil insecticide

(c)   Fargo Ltd.

Exercise 21.7. Cutworms

Results

1. (a) Caterpillars are ground level biting/chewing feeders.

    (b) None.

    (c) Occasionally troublesome.

2. 35–50 mm.

3. Soil colour brown.

4. 3 pairs of true legs and five pairs of false legs.

Conclusions

(a)   chlorpyrifos

(b)   Suscon Green soil Insecticide

(c)   Fargo Ltd.

Exercise 21.8. Common gooseberry sawfly (*Nematus ribesii*)

Results

1. (a) Caterpillars are biters/chewers feeding mainly on leaves.

    (b) None.

    (c) Very occassional sucking damage.

2. Two pairs.

3. Black head with a green body with black spots.

4. 30–35 mm

Conclusions

For example:

(a)   cypermethrin

(b)   Toppel 10

(c)   United Phosporus Ltd.

Exercise 21.9. Cabbage white butterfly *(Mamestra brassicae)*

Results

1. (a) caterpillars attack aerial parts by biting/chewing.

   (b) None.

   (c) None.

2. May/June on the underside of leaves.

3. Up to 40 mm.

4. Bluish to yellow-green with a yellow line along the back.

5. September/October.

Conclusions

For example:

(a) cypermethrin

(b) Toppel 10

(c) United Phosporus Ltd.

Exercise 21.10. Red spider mite *(Tetranychus urticae)*

Results

1. 6 legs

2. 0.5 mm

3. (a), (b) and (c): the spider mite does not have a larvae or pupa stage; instead grows larger from a juvenile to adult mite. All stages damage the plant as sap-sucking insects.

Conclusions

1. For example:

   (a) tebufenpyrad

   (b) Masai

   (c) BASF plc.

2. *Phytoseiulus persimilis* predatory mite

3. Mating

Exercise 21.11. Aphids *(Hemiptera)*

Results

1. 6 legs

2. 1.4–2.6 mm

3. (a), (b) and (c): aphids undergo incomplete metamorphosis, getting bigger as they get older, often shedding their skin along the way. Both juvenile and adults are sap-suckers, usually feeding on aerial plant parts.

Conclusions

For example:

(a) imidacloprid

(b) Intercept 5GR

(c) Scotts Company.

Chapter 22   Nematodes

Key facts

1. Nematodes are parasites of other animals and plants that live in the soil.
2. Just a teaspoon of soil contains thousands of nematodes and their global population is more than any other animal with over 20 000 species classified.
3. Nematodes spread quickly, being carried on boots, tools and machinery.
4. Some nematodes can be used to control other pests as a method of biological control.
5. Females swell up into a round cyst containing 200–600 eggs.
6. Good plantsmanship and rotation is the key to effective control.

Background

Nematodes, also called eelworms and roundworms, due to their appearance, are plant pests that live in the soil and are normally encouraged by wet conditions. They help in the breakdown and recycling of organic matter in the soil to form humus.

Nematodes are unsegmented animals and very different from ordinary worms. They have long thin thread-like bodies ('nema' is Greek for 'thread'). They are parasites of insects, plants and animals. The largest worms are thought to live in Australia and can grow up to 3 m long. However, the largest nematode ever observed (*Placentonema gigantisma*), was 8 m long and discovered in a sperm whale. The *Ascaris* nematode lives in the intestines of pigs and grows up to 25 cm long. It is the largest nematode found in Britain, and may be deadly to man. Many are small or microscopic and occur in extremely large numbers, especially in wet conditions. There are over 20 000 species of nematodes classified worldwide. They are the most numerous animals on the planet: just a teaspoon of soil will contain thousands of nematodes.

There are many types of nematodes that effect plants in the UK including:

> beet cyst nematode (*Heterodera schachtii*)
> pea cyst nematodes (*Heterodera goettingiana*)
> cereal cyst nematode (*Heterodera avenae*)
> chrysanthemum nematode (*Aphelenchoides ritzemabosi*)
> leaf nematodes (*Aphelenchoides fragariae* and *Aphelenchoides ritzemabosi*)
> stubby-root nematode (*Trichodorus* and *Paratrichodous* spp)
> needle nematodes (*Longidorous* spp)
> stem nematodes (*Ditylenchus dipsaci*)
> root-lesion nematodes (*Pratylenchus penetrans*)
> dagger nematodes (*Xiphinema diversicaudatum*)
> potato cyst nematode (*Globodera rostochiensis* and *Globodera pallida*)
> potato tuber nematode (*Ditylenchus destructor*)
> root knot nematodes (*Meloidogyne* spp.).

These affect a vast range of plants from protected crops including tulips, strawberries, vegetables and some weeds. The main cause of infestation is growing a plant too frequently in the rotation and wet conditions. Several species of nematodes are parasites of garden pests such as slugs, and several commercial products are now available for purchase as a biological control agent (see Chapter 23).

Common characteristics

Effects on plants

Low populations only damage a few roots and horticulturalist may be unaware that the land is infected. As nematode populations increases so plant yields and quality decrease.

Symptoms of attack

- small patch or patches in the crop
- stunted growth and unhealthy looking foliage
- outer leaves wilt in sunshine, turn yellow prematurely and die
- the heart leaves may remain green but are undersized
- roots remain stunted
- plants are easily removed from the soil
- tap root is small and short or absent
- development of 'hunger' roots (excessive lateral root development)
- small lemon-shaped nematodes can be seen protruding from dead roots.

Method of spread

- soil movements, e.g. on tractors/machinery/animals
- carried on plants, e.g. seed potatoes/plants for transplanting
- strong winds blowing soil
- too frequent cropping with hosts plants.

Control methods

- use resistant plant varieties
- biological control e.g. parasitic fungi
- crop rotations may be the only economical method of control e.g. crop only 1 year in three as a minimum, rising to one in five on poor soils.

Exercise 22.1

Vinegar nematode (*Angiullula aceti*)

Background

Nematodes are simple organisms containing a tube-like gut and separate mouth and anus (see Figure 22.1). Typically they may be seen as small, white or transparent threads of cotton wool. They live by their millions in almost every habitat. It has been estimated that 1m² of woodland soil contains ten million nematodes. They move through the soil in films of water in a characteristic

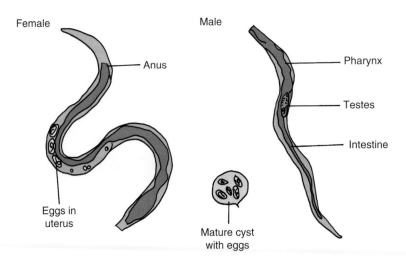

Female

Male

Anus

Pharynx

Testes

Intestine

Eggs in uterus

Mature cyst with eggs

Figure 22.1 Typical nematode biology

'side-winder'-like motion caused by the contraction of four muscles which bend and twist the body.

The vinegar nematode causes the fermentation and characteristic acid taste to vinegar and old beer. Upset tummies resulting from drinking outdated cloudy beer are sometimes caused through nematodes being passed into the stomach. They are an easy source of nematodes to collect and a useful starting point to aid identification.

Aim
To observe living male and female nematodes.

Apparatus
Garden centre purchased nematodes kit or infected cider vinegar/cloudy beer
Microscope/hand lens
0.2% neutral red stain
Filter paper
Slides and cover slip
Dropping pipette

Useful websites
www.nematodes.org

www.fotosearch.com/CRT778/001205cf/

Method
1. Add a drop of infected water solution to a microscope slide.
2. If desired, add one droplet of 0.2% neutral red stain to the solution to aid clarity.
3. Observe the organisms under the microscope and note the speed with which they move.
4. Place a cover slip over the solution.
5. Place the filter paper against the edge of the cover slip to withdraw some water and slow down the movement of the nematodes.
6. Observe the different growth stages present.
7. State how many male and females are present.
8. What are the two main differences between the male and female nematodes?
9. How large is each nematode (estimate): 0.25 mm, 0.5 mm, 1 mm, 2 mm or 3 mm?

Results
Enter your observations in the table provided.

Features of vinegar nematodes			
Estimate of sizes of nematodes present		Differences between males and females	Growth stages present
Number of males		1 2	1 2
Number of females		3	3

Conclusions

1. Explain why some people experience an upset tummy after drinking real ale.
2. State a positive use of nematodes in horticulture.
3. Describe the symptoms of nematode attack on horticultural plants.

Exercise 22.2

Nematode damage to plants

Background

Nematode damage can remain undetected until severe plant losses occur. A list of the typical symptoms is given in the earlier background section and may be helpful to refer to here. A good source of materials are rotten bulbs and plants that have been thrown onto a compost heap.

Aim

To recognize symptoms of nematode attack in a range of plants.

Apparatus

Root knot nematode on cucumber and carnations
Chrysanthemum nematode on chrysanthemums
Narcissus stem bulb nematode
Stem and bulb nematode on onion

Useful websites

www.rothamsted.ac.uk/broom/clinic/otherrootprobs.php

www.defra.gov.uk/planth/pestnote/meloid.htm

www.defra.gov.uk/planth/qic.htm

http://services.csl.gov.uk/cut_flowers.cfm

Method

Collect specimens of the above-listed plants and describe the damage caused in each.

Results

Enter your observations in the table provided.

Nematode	Symptoms of damage
Root knot nematode (*Meloidogne* spp) 1. Cucumber 2. Carnation	
Chrysanthemum nematode (*Aphelenchoides ritzemabosi*) 1. Chrysanthemum	
Stem and bulb nematode (*Ditylenchus dipsaci*) 1. Narcissus 2. Onion	

Conclusions

1. State how each of the above crop pests may be controlled or prevented.
2. List typical methods by which nematodes are spread.
3. Explain what is meant by the term 'parasitic nematode'.

Exercise 22.3

Potato cyst nematode (*Globodera* spp) in soils

Background

> Potato cyst nematodes can be found in most soils, but particularly those which have been used to grow potatoes.

They result in damaged and stunted plants with patchy growth and low yields. The nematodes burrow into the plant and after fertilization the female swells up into a hard ball containing between 200–600 eggs. This egg case is called a 'cyst'. They can survive in the soil for up to 10 years and are easily identified by their red-brown lemon-shaped pods. Sometimes they can be confused with weed seeds. Figure 22.1 shows the biology of nematodes. Once the population size of this nematode has been determined an appropriate management practice can be selected and implemented to control the infestation.

In this exercise a sample of soil is mixed with water. The soil mineral fraction sinks as it is heavier than the cysts which will remain floating, along with other floating material, and is poured out (decanted) onto filter paper so that they can be counted.

Aim

To ascertain the population size of potato cyst nematode cysts in a soil sample.

Apparatus

50 g air dry soil	1 litre conical flask
Large funnel	Tripod
Filter paper	White tray
Hand lens	Balance
Stirring rod	Weigh boat
Beaker	

Useful websites

www.rhs.org.uk/advice/profiles0801/eelworm.asp

www.defra.gov.uk/planth/pestpics/qic2004/QIC65.pdf

www.niab.com/services/laboratory-services/potato-cyst-nematodes.html

www.invasive.org/browse/subimages.cfm?sub = 4905

Method

1. Weigh out approximately 50 g of air dried soil ground to pass a 2 mm sieve.
2. Transfer the soil to a 1 litre conical flask.
3. Make up to the 400 ml mark with tap water.
4. Swirl to thoroughly wet-up (hydrate) the soil.
5. Fill flask to the top using a trickle of water so that no floating material is spilt.

6. Fold the filter paper into four and place in the funnel in the tripod over the beaker (see Figure 22.2).
7. Pour the floating material into the funnel, simultaneously rotating the flask.
8. After decanting 5 cm of solution, refill the flask and continue as in 7 until all the floating material has been decanted into the funnel.
9. Use the stirring rod to swirl the solution retained by the filter paper and encourage the kitenous material to move onto the filter paper.
10. After five minutes pierce the bottom of the filter paper and allow to drain.
11. Empty waste soil remaining in flask into waste bin.
12. Transfer the filter paper onto a white tray.
13. Use a hand lens to count the number of cysts present (red-brown lemon-shaped pods) (see Figure 22.1).
14. Do not confuse weed seeds with cysts.

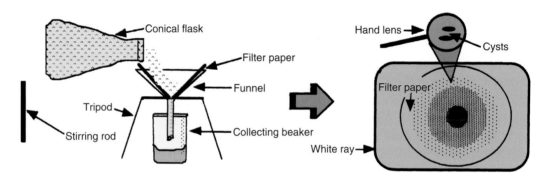

Figure 22.2 Apparatus for potato cyst nematode extraction

Results

State your results in the table below and select your management practice.

Number of cysts found per 50 g in soil sample=	
Number of cysts per 50 g	Management practice
0	Plant potatoes of any variety
1–5	Probably will get good crop with any variety
5–10	May get losses in patches with susceptible varieties
11–20	Probably all right with earlies or plant a resistant variety or use a soil nematicide
21–40	Leave land for 2–4 years
≥40	Leave land for 6 years

Resulting management decision for sample soil =

Conclusions

1. How may the pest damage be recognized by looking in the field at growing plants?
2. How may the pest damage be recognized by looking at the roots?
3. How does the pest survive in the soil?
4. How many eggs may a cyst contain?
5. What has the cyst stage developed from?
6. How long may eggs survive in the soil?
7. How may a cyst be recognized with a hand lens?

Answers

Exercise 22.1. Vinegar nematode (*Angiullula aceti*)

Results

Males and females should be counted and the differences between them described. Figure 22.1 will be a useful guide.

Conclusions

1. The real ale may be infected by vinegar nematodes.

2. Some species can be used as biological controls, e.g. *Phasmarhabditis hermaphrodita* to control slugs (see Chapter 23).

3. Typical symptoms include: small patch or patches in the crop, stunted growth and unhealthy looking foliage, outer leaves wilt in sunshine, turn yellow prematurely and die, the heart leaves may remain green but are undersized, roots remain stunted, plants are easily removed from the soil, tap root is small and short or absent, development of 'hunger' roots (excessive lateral root development), small lemon-shaped nematodes can be seen protruding from dead roots.

Exercise 22.2. Nematode damage to plants

Results

Root knot nematodes: knarled and warped roots, presnce of nodules on roots.
Chyrsanthemum nematode: brown leaf tips, leaves shriveled.
Stem and bulb nematode: black bulb heart, leaves brown and turned up.

Conclusions

1. Prevent by good hygiene and appropriate crop rotation. Control by appropriate chemical such

as containing *1,3-dichloropropene* in 'Telone II' manufactuered by Dow AgroSciences.

2. For example: soil movements e.g. on tractors/machinery/animals, carried on plants e.g. seed potatoes/plants for transplanting, strong winds blowing soil, too frequent cropping with hosts plants.

3. Deriving their nutition by feeding off a living organism.

Exercise 22.3. Potato cyst nematode (*Globodera* spp)

Results

Appropriate management practice should be selected from the table.

Conclusions

1. Patchy growth, low yield.

2. Roots are damaged and stunted growth.

3. As a cyst.

4. 200–600 eggs.

5. A swollen fertilized female nematode.

6. Up to ten years.

7. Red-brown lemon-shaped pod.

Chapter 23   Biological control

Background

Figure 23.1 Ladybirds often used as biological controls of aphids

The principle of biological control is the use of beneficial organisms (natural enemies) as **pathogens**, **predators** or **parasites** to control common pests as an alternative to using chemicals.

A predator is simply an organism that consumes another providing energy, such as a fox eating a chicken. They may kill many pests during their life cycle. For example the mite called *Phytoseiulus persimilis* hunts and kills glass house red spider mite. Probably the best known biological control using predators is the practice of introducing ladybirds to a garden to feed on the greenfly population. Another type of tropical ladybird called *Cryptolaemus montruzieri* (Figure 23.2) has been used very successfully, in the control of mealybugs.

Figure 23.2 *Cryptolaemus montruzieri* adult feeding on mealybugs (courtesy Koppert Biological Systems)

Figure 23.3 *Metaphycus helvolus* parasitizing scale insect (courtesy Dr Mike Copland, Wye College)

A parasite is an organism that feeds off of a living host pest but, in biological control, also results in the eventual death of the pest. Mostly, only one pest is killed during the life cycle of the parasite. Examples include the use of a parasitic wasp called *Metaphycus helvolus* (Figure 23.3) which lays its eggs in the juvenile stage of scale insects. As the wasp egg hatches the grub eats away at the protein-rich scale, killing the pest.

Similarly, consider the increasing use of nematodes (Figure 23.4) to control garden slugs and vine weevils. The nematode burrows into the slug's skin and eats it from within. In addition, crushed egg shells can be used to create an uncomfortable surface over which slugs will not want to crawl, to protect plants, creating a barrier to discourage slugs without using chemicals.

A pathogen, by comparison, is normally a form of disease; examples include the use of bacteria and fungus to kill pest insects.

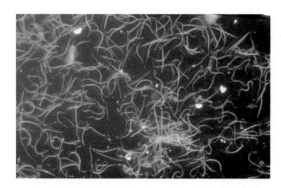

Figure 23.4 *Heterorhabditis megidis* nematodes (courtesy Koppert Biological Systems)

The use of biological control agents (natural enemies) is increasing

Several pests are 'notifiable' and if found should be reported to the Ministry of Agriculture, Fisheries and Food. Examples include Colorado beetle, plum pox, South American leaf miner, western flower thrip and wart disease on potatoes. However, many natural enemies have been introduced to Britain from overseas. *Phytoseiulus*, for example, was first discovered in Germany in the 1960s on a consignment of orchids arriving from Chile.

Many pests have grown immune/resistant to chemicals and it is a continuing contest to create effective pesticides. Every time a pesticide is used, the plant's quality (colour, texture, consistency etc.) has been shown to decrease. Approximately 1–2 per cent of yield may be lost every time chemicals are used. Biological control is now the standard pest control on salads grown under glass. There is also growing pressure against the use of chemicals in other food crops. For example, advice in dry summers has been to peel carrots as some of the outer skin may contain harmful pesticide residues. In addition pesticides and herbicides have begun to turn up in bore-hole water supplies at similar concentrations to nitrates, before it was realized that nitrates were a problem. Much research is being conducted to extend our knowledge in these areas and at reducing our dependence upon chemical control. Many glasshouse crops are grown under contract to supermarkets who stipulate that no chemical controls are to be used without their prior approval. 'Nature's Choice'

Figure 23.5 Crushed eggshells make an effective barrier to slugs and all the family can join in

and organic produce is sold at a premium price because, as far as practicable, they have been grown without using chemicals. Pot plants and cut flowers, however, present more of a problem. Ideally, control by biological methods is desirable but the public are wary about buying plants with 'strange' insects (predators) crawling around on them. There is still a great deal of consumer educating still to be done in this field.

Advantages of biological control	Disadvantages of biological control
• Environmental benefits	• Initially slow to establish and gain control
• Increase in yield and quality	• Live organisms that sometimes die in transport
• No halt to plant growth (phytotoxicity)	• Operators require knowledge of life cycle stages
• Permanent solution to pest problem	• Pesticide and insecticides use restricted
• Specific to pest	• Correct environmental conditions are needed (e.g. light and humidity)
• Generally safe to other beneficial organisms Safer to operators and public	• May consume other natural enemies (e.g. Macrolophus eating Phytoseiulus)
• No 'resistance' problems	• Less successful on outdoor plants due to dispersal and colder temperature
	• Life forms are unpredictable and may be difficult to rear

In the exercises in this chapter, some aspects of biological control will be investigated as an introduction to an increasingly important area of horticulture. Example predators can be purchased from suppliers listed in the back of many trade magazines and in the section titled 'Suppliers' at the end of this book. Alternatively, most commercial horticulturalists and increasing numbers of garden centres stock some natural controls or can order biological control agents.

Whitefly (*Trialeurodes vaporariorum*) case study

The glasshouse whitefly is a tiny, white moth-like insects of Asian origin, from the order Hemiptera. They are related to aphids, having very similar mouth parts. They are different in that they have a scale stage and always lay eggs. The cabbage whitefly (*Aleyrodes proletella*) attacks brassicas and related plants outdoors.

Whitefly feed by the piercing and sucking of leaf and petal tissues, mainly by the larvae stage of the life cycle (see Figure 23.6) which stick to the lower epidermis and suck the sap. The larvae also excrete honeydew which can encourage sooty mould.

Life cycle

The whitefly has an incomplete metamorphosis life cycle of five stages: one egg, three larval, one adult. Eggs are laid on the underside of leaves in circles. After a few days the eggs change in colour from yellow-white

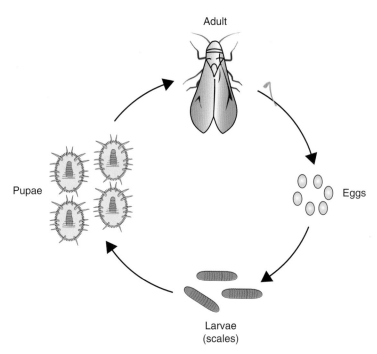

Figure 23.6 Whitefly life cycle

to black and hatch. The first emerging nymphs (instars) are almost invisible. They crawl a few centimetres then settle for life.

After a further week the scale stage develops. These 0.5 mm larval nymphs are pale-green flat scales, not mini adults as with other Hemipterans, and feed by stylets like aphids. Including the crawlers, four stages of scale exist; the last is equivalent to a thick walled pupa. They gradually increase in size with each moulting until pupation when they look like a raised oval disk. The pupae are white, hairy and thick skinned. The emerging adults are white, due to a mealy wax, 1 mm long and live for four weeks. They reproduce without mating (parthenogenic) and start laying after three days, at a rate of two hundred eggs per adult.

Typical plants favoured by whitefly include; Begonia, Calceolarias, Cinerarias, Chrysanthemum, Coleus, Dahlias, Fuchsia, Freesias, Gerbera, Heliotropes, Hibiscus, Impatiens, Pelargonium, Primulas, Poinsettia, Salvias, Solanum, Verbenas, aubergine, tomatoes, cucumbers, peppers and French beans.

Symptoms of whitefly attack include: weak growth, sticky honey dew, sooty mould growth, yellow leaves, loss of vigour and annoyance at cloud plumes (snow showers) of whiteflies.

Control methods

Cultural control

Monitor/sample using sticky traps. Remove weeds which may act as an alternate host and inspect plants to ensure that none is already infected.

Chemical control

Some insecticides used are alginate, pyrethrins and malathion.

Biological control

The best known biological control is the use of a parasitic chalcid wasp, 0.6–1 mm in size, called *Encarsia formosa*. The adult wasp attacks whitefly larvae scales, stinging them and laying an egg in the larvae body. Excessive stinging will kill the larvae without the need for parasitization with eggs. The life cycle interaction is shown in Figure 23.7.

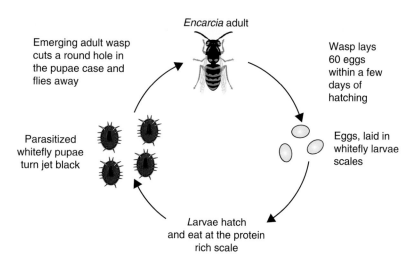

Figure 23.7 *Encarsia formosa* control of whitefly

Encarsia is introduced weekly, as black parasitized whitefly scales on a card which is hung on the plant. The best control is achieved by a 'little and often' introduction of the wasp. This is sometimes called the dribble method; examples include using 1 wasp/plant at each stage of introduction and 5 introductions at 7–10 day intervals. For ornamentals apply 5–10 parasites/m^2, when the density of whitefly is no more than 1 per 2 m^2.

Exercise 23.1

Common uses of biological control agents

Background

Predators and parasites control pests in distinctly different ways.

The following biological control agents include both types, but should be considered with reference to their practical use in horticulture. For example, the use of the parasitic wasp *Encarsia formosa* is only effective if introduced early at a density of one wasp scale per plant, when the whitefly pest problem is no greater than one whitefly per twenty plants. They also requires a minimum temperature of 20°C, otherwise they do not fly around from plant to plant and walking gives less effective control.

Tomato growers also benefit from a predator which can consume between ten and fifty whitefly per day. It is called *Macrolophus caliginosus* and has been used successfully in Europe. They have also been known to feed on spider mites, moth eggs and aphids. Used together with *Encarsia*, they can control whitefly more effectively. Obviously, no insecticides should be used as they are likely to kill the agent. Table 23.1 summarizes the breadth and type of natural enemy that can be used to control a range of insect pests.

Table 23.1

Pest	Natural enemies	
Aphids	*Aphelinus abdominalis*	– parasitic wasp
	Aphidius colemani	– parasitic wasp
	Aphidoletes aphidimyza	– gall-midge
	Chysoperla carnea	– lacewing
	Delphastus pusillus	– predatory beetle
	Hippodamia convergens	– ladybird
	Verticillium lecanii	– pathogenic fungus
	Metarhizium anisopliae	– pathogenic fungus
Caterpillars	*Bacillus thuringiensis Trichogramma evanescens*	– pathogenic bacterium extract
		– parasitic wasp
		– pheromone traps
Leaf-miners	*Diglyphus isaea*	– parasitic wasp
	Dacnusa sibirica	– parasitic wasp
Mealybugs	*Cryptolaemus montruzieri*	– predatory tropical ladybird
	Leptomastix dactylopii	– parasitic wasp
Red spider mite	*Phytoseiulus persimilis*	– predatory mite
	Therodiplosis persicae	– predatory mite
	Typhlodromus pyri	– predatory mite
	Amblyseius californicus	– predatory mite
Leaf hopper	*Anagrus atomus*	– parasitic wasp
Scale insects	*Metaphycus helvolus*	– parasitic wasp
	Chilocorus	– predatory beetle
Sciarid flies	*Steinernema carpocapsae*	– parasitic nematode
	Hypoaspis miles	– predatory mite
Slugs	*Phasmarhabditis hermaphrodita*	– parasitic nematode
Thrips	*Amblyseius mackenziei , A. barkeri*	– predatory mites
	A.cucumeris, A. degenerans	– predatory mites
	A. Orius majusculus, O. Spp.	– predatory bugs
	Verticillium lecanii	– pathogenic fungus
Vine weevils	*Steinernema carpocapsae*	– parasitic nematode
	Heterorhabditis megidis	– parasitic nematode
Whiteflies	*Encarsia Formosa*	– parasitic wasp
	Macrolophus caliginosus	– predatory bug
	Delphastus pusillus	– predatory black beetle
	Verticillium lecanii	– pathogenic fungus

Aim

To recognize beneficial insects and their method of controlling undesirable pests.

Apparatus

Various agents including: *Orius laevigatus, Bacillus thuringiensis, Verticillium lecanii* parasitized whiteflies, *Phytoseiulus persimilis* mite, nematode parasitized slugs and vine weevils, *Encarsia formosa* parasitized whitefly scales

Microscopes

Variety of hand lenses to test.

Useful websites

www.koppert.com

www.defenders.co.uk

www.allotment.org.uk

www.nvsuk.org.uk/garden_equipment

Method

1. Observe the biological control agent under the microscopes.
2. Describe your observations (e.g. diagram showing size and characteristics).
3. State which pest the natural enemy controls and whether predator, parasite or pathogen.
4. Repeat for other agents.

Results

1. Whitefly scales parasitized by the *Encarsia formosa* wasp (see Figure 23.8).
   (a) Decribe it (e.g. shape, colour, legs, antennae, speed of movement/flight).
   (b) Is this a pathogen, parasite or a predator?
   (c) List five other wasps and the pests that they control.
2. *Orius laevigatus*, an anthocorid bug (see Figure 23.9) (also investigate *O. insidiosus, O. tristicolor, O. albidipennis*, and *O. majusculus*).
   (a) Decribe it (e.g. shape, colour, legs, antennae, speed of movement/flight).
   (b) Is this a pathogen, parasite or a predator?
   (c) Which pest does it control?
   (d) What additional action should be taken if this particular pest is discovered on your plants?
   (e) List one other bug and the pest that it controls.
3. *Phytoseiulus persimilis* mite (see Figure 23.10).
   (a) Decribe it (e.g. shape, colour, legs, antennae, speed of movement/flight).
   (b) Is this a pathogen, parasite or a predator?
   (c) Which pest does it control?
   (d) List five other mites and the pests that they control.
4. Nematode (*Phasmarhabditis hermaphrodita*) parasitized slugs (see Figures. 23.11 and 23.12) and vine weevils.
   (a) Decribe it (e.g. shape, colour, legs, antennae, speed of movement/flight).

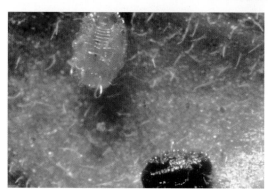

Figure 23.8 Whitefly scale parasitized (black) and unparasitized (white) by parasitic wasp *Encarsia Formosa* (courtesy Koppert Biological Systems)

Figure 23.9 *Orius laevigatus* adult attacking adult thrip (courtesy Koppert Biological Systems)

Figure 23.10 *Phytoseiulus persimilis* attacking red spider mite (courtesy Koppert Biological Systems)

Figure 23.11 Slug with swollen mantle after infection with nematodes (courtesy Defenders Ltd)

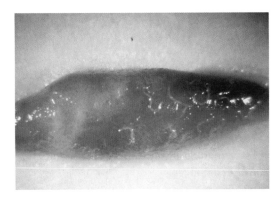

Figure 23.12 Slug killed and decomposing to release another generation of parasitic nematodes (courtesy Defenders Ltd)

Figure 23.13 Whitefly adult and larvae infected by *Verticillium lecanii*

Figure 23.14 Caterpillar killed by *Bacillus thuringiensis*, suspended from leaf by hind legs (courtesy Koppert Biological Systems)

    (b) Is this a parasite or a predator?

    (c) Which pest does it control?

    (d) List two other species of nematode and the pests that they may control.

5. *Verticillium lecanii* fungus infected whiteflies (Figure 23.13).

    (a) Describe it (e.g. shape, colour, legs, movement/flight, fungus on legs, hyphae visible, mouldy, fluffy).

    (b) Is it a pathogen, parasite or a predator?

    (c) Which pest does it control?

    (d) List two other pests that *Verticillium lecanii* may control.

6. *Bacillus thuringiensis* (a bacterium extract) infected caterpillars (see Figure 23.14).

    (a) Describe it (e.g. shape, colour, legs, movement, paralyzed, general appearance).

    (b) Is it a pathogen, parasite or a predator?

    (c) Which pest does it control?

    (d) List two other biological controls for caterpillar pests.

Conclusion

1. State the meaning of the term biological control.

2. Describe four precautions that need to be taken to ensure effective parasitism of whitefly by *Encarsia* wasp.

3. Explain the term 'parasite' as used in biological control.

4. Give an example of a parasite other than *Encarsia*, and the insect that it attacks.
5. Explain the term 'predator' as used in biological control.
6. Give an example of a predator and the pest that it preys on.
7. Give two examples of biological control of pests in horticulture.
8. Describe four other control organisms for use against named glasshouse or field pests.
9. Explain why the use of biological control agents is likely to increase.

Exercise 23.2

Diagnosis of pest damage

Background

Diagnosing plant disorders can be a challenging occupation. Unhealthy plants sometimes give rise to conditions that favour pest attack. Once a pest becomes established on a plant it may invite fungal diseases, feeding on honeydew and excrement, or viruses, introduced through the method of feeding. One thrip can infect fifty tomato plants with tomato spotted wilt virus (TSWV).

When the pest has been identified a suitable control may be selected. An essential aid to this section should be the purchase of a small pocket hand lens ($\times$10 magnification) or similar magnifying glass. This can be used to identify both pests and symptoms of plant damage.

Problem pests can normally be separated into two groups: those that damage the plant by biting/chewing and those that are sap-suckers (see Chapter 21). Earwigs, caterpillars, slugs and vine weevils, for example, all chew the foliage (Figure 23.15).

The sap-suckers, including aphids and thrips, pierce the leaf and petals with their needle-like stylet and then suck out the nutrient-rich sap, causing air to rush into the cells, collapsing them. In the case of thrips this leaves behind a characteristic silver/white streaking pattern (Figure 23.16).

Leaf-miners may be recognized by the piercing dots left on the leaves. Similarly, snails may be recognized by the slime trail left behind. Figure 23.17 is a field guide of typical symptoms of plant damage by insect pests which may be a useful aid to describing the visual signs of pest damage.

Aim

To describe the visual symptoms of a range of pest damage

Apparatus

Hand lens
Selection of problem plants
Figure 23.17

Method

1. Observe some problem plants attacked by a range of insect pests.
2. Describe the symptoms of the damage caused, using the field guide (Figure 23.17) as a pointer.

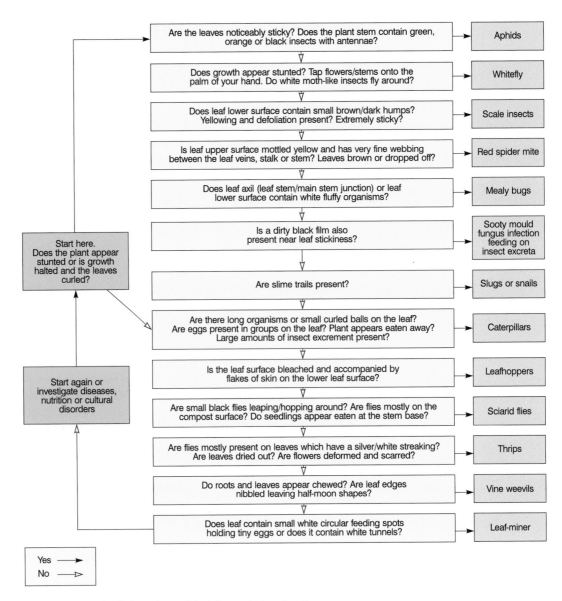

Are the leaves noticeably sticky? Does the plant stem contain green, orange or black insects with antennae?	Aphids
Does growth appear stunted? Tap flowers/stems onto the palm of your hand. Do white moth-like insects fly around?	Whitefly
Does leaf lower surface contain small brown/dark humps? Yellowing and defoliation present? Extremely sticky?	Scale insects
Is leaf upper surface mottled yellow and has very fine webbing between the leaf veins, stalk or stem? Leaves brown or dropped off?	Red spider mite
Does leaf axil (leaf stem/main stem junction) or leaf lower surface contain white fluffy organisms?	Mealy bugs
Is a dirty black film also present near leaf stickiness?	Sooty mould fungus infection feeding on insect excreta
Are slime trails present?	Slugs or snails
Are there long organisms or small curled balls on the leaf? Are eggs present in groups on the leaf? Plant appears eaten away? Large amounts of insect excrement present?	Caterpillars
Is the leaf surface bleached and accompanied by flakes of skin on the lower leaf surface?	Leafhoppers
Are small black flies leaping/hopping around? Are flies mostly on the compost surface? Do seedlings appear eaten at the stem base?	Sciarid flies
Are flies mostly present on leaves which have a silver/white streaking? Are leaves dried out? Are flowers deformed and scarred?	Thrips
Do roots and leaves appear chewed? Are leaf edges nibbled leaving half-moon shapes?	Vine weevils
Does leaf contain small white circular feeding spots holding tiny eggs or does it contain white tunnels?	Leaf-miner

Start here. Does the plant appear stunted or is growth halted and the leaves curled?

Start again or investigate diseases, nutrition or cultural disorders

Yes ⟶
No ⟶

Figure 23.15 Field guide of typical symptoms of plant damage by insect pests

Figure 23.16 Biting and chewing slug damage

Figure 23.17 Sap-sucking insect damage to petals

Results

Produce a table of results showing the types of symptoms found against the insect pest investigated.

Conclusions

1. Describe the different symptoms of pest attack between plants that damage by sucking and biting.
2. Why are unhealthy plants more likely to attract pest insects than healthy plants?
3. Describe why fungus diseases may become easily established on plants suffering from pest attack.

Exercise 23.3

Monitoring pest levels

Background

An essential aid to correct pest management is the use of monitoring devices to correctly indicate the population size of the problem pest. In short, detection is the basis of control. The sooner harmful organisms are detected, the more effective will be a biocontrol programme.

Locating the first pest insect can be difficult; 'trapping' can be a very important aid for this. For outdoor grown plants the use of a pit fall trap, as described in Chapter 21, is highly recommended. For glasshouse-grown plants the use of sticky traps are essential. These are advanced forms of fly paper to which the insects stick and can be subsequently identified and counted. There should be no such thing as an empty glasshouse; a number of pests are able to survive the period between crops and may be caught before they can damage new plants, thereby helping to make a clean start to the cropping programme. Glasshouses should always have sticky traps present by the doors and vents so that an accurate assessment can be made of pest levels through constant monitoring.

There are several traps available including white, orange, yellow and blue. These colours appear to be more attractive to insects, although other factors, such as glue type, size and shape have also been found to be important. White and orange traps are used for leaf miner monitoring. The most extensively used traps are blue and yellow. The yellow traps are the most common and slightly cheaper, used to attract whitefly, leaf miners, aphids and sciarid flies. The effectiveness of the traps are influenced by environmental factors including day length, light radiation and temperature, which effect the behaviour pattern of the pests insects.

The blue traps are essential to detect thrip levels as thrips are not easily attracted to the yellow traps. The blue traps catch up to four times as many female thrips as the the yellow ones. Counts as low as 4–5 thrips per sticky trap should be considered dangerous levels. Chemical sprays may be needed to lower pest levels prior to the introduction of a natural enemy.

Three simple principles should govern the use of these biological agents:

1. Introduce biological agents early in the season before pest levels rise.
2. Introduce biological agents at an appropriate density to control the pest (e.g. *Amblyseius cucumeris* at 50 predators per m² per week all year round to control thrips).

3. Adopt a whole site approach. Good husbandry in one glasshouse may be counterproductive if pests can fly in through vents from another badly infected house.

There are instances where sticky traps may be used as a control. However, as a general rule, sticky traps are not a pest control method. They are a management tool useful to accurately assess pest population levels and devise suitable control programmes based upon the information that they yield.

Natural enemies are readily available to commercial growers and can be cheaper than chemicals because pests are increasingly immune. The use of biological agents themselves does not always run smoothly. Being live organisms they sometimes die in transport and do require the horticulturalist to have greater knowledge of the life cycles of both the agents and the pests than might otherwise be the case with chemical control. Generally horticulturalist should approach the subject through a balance of pest management methods to ensure effective control.

Aim

To use a range of techniques to detect and monitor pest insects.

Apparatus

blue and yellow sticky traps
any insect identification key/handbook

Useful websites

www.defenders.co.uk

www.syngenta-bioline.co.uk/productdocs/html/StickyTraps.htm

Method

1. Set up a set of blue and yellow sticky traps in different glasshouses.
2. Observe the traps weekly and record the insect pests caught over the next three weeks.

Results

Enter your results in the table provided.

Observation period	Yellow sticky trap	Blue sticky trap
Week 1		
Insect pests detected		
Number of occurrences		
Week 2		
Insect pests detected		
Number of occurrences		
Week 3		
Insect pests detected		
Number of occurrences		

Conclusions

1. List five difficulties associated with biological control and explain why its use is likely to increase.
2. Describe the application of biological control organisms to a named crop with which you are familiar.
3. Name two organisms commonly used for biological control in glasshouse flower crops.
4. Explain the importance of pest monitoring and describe some methods to accumulate adequate information.
5. State the advantage of using blue traps rather than yellow traps for thrip monitoring.
6. Describe the environmental factors which influence the successfulness of using sticky traps.
7. For each of the insect species detected on the sticky traps, state a suitable beneficial organism which may be used as biological control agent.

Exercise 23.4

Predators and prey

Background

Pest management is a continuing battle of selecting the right weapon from the armoury to defeat the organism.

Predators will often only lay eggs when pest numbers are causing plant losses. This is to ensure that there is sufficient prey for the hatched young to feed on. Ladybird numbers, for example, correspond closely with the population size of lacewings.

Certain biological control agents feed on more than one form of prey and are called polyphagous. Parasitic nematode strains consume not only slugs but also sciarid fly larvae and will attack a range of insects in the right conditions. Similarly, strains of *Verticillium lecanii* infect both whiteflies, thrips and aphids. Sometimes both chemicals and biological agents may work together to defeat a range of insects. This is referred to as integrated pest management. In this exercise you are asked to select a biological weapon for a named horticultural pest.

Aim

To apply horticultural knowledge in matching biological control predators to pests.

Apparatus

List of natural enemy predators and prey
Knowledge from previous exercises

Method

1. Match the following list of natural enemies to the pest problems listed.

Natural enemies	Pests
Bacillus thuringiensis bacterium	Whiteflies
Anagrus wasp	
Aphelinus abdominalis wasp	
Aphidius colemani wasp	Sciarid flies
Diglyphus isaea: wasp	
Dacnusa sibirica wasp	
Encarsia Formosa wasp	Leaf-hopper
Leptomastix dactylopii wasp	
Metaphycus helvolus wasp	Mealybugs
Trichogramma evanescens wasp	
Aphidoletes aphidimyza gall-midge	
Chysoperla carnea lacewing	Caterpillars
Chilocorus beetle	
Delphastus pusillus beetle	Thrips
Cryptolaemus montruzieri tropical ladybird	
Hippodamia convergens ladybird	Aphids
Amblyseius mackenziei , A. barkeri mites	
A. cucumeris, A. degenerans mites	Slugs
Hypoaspis miles mite	
Phytoseiulus persimilis mite	
Therodiplosis persicae mite	Red spider mite
Typhlodromus pyri mite	
Macrolophus caliginosus bug	Vine Weevil
Orius majusculus, O. spp. bugs	
Steinernema carpocapsae nematode	
Steinernema bibionis nematode	Leaf-miner
Heterorhabditis megidis nematode	
Phasmarhabditis hermaphrodita nematode	
Verticillium lecanii fungus strains	Scale insects

2. Use a thick line for main prey, and a dotted line for secondary prey, for example, Verticillium lecanii has aphids as a main prey, and thrips and whitefly as a secondary prey.

Results

Indicate which pests are the main and secondary prey for each natural enemy.

Conclusions

1. Explain the term 'predator' as used in biological control.
2. Describe the integration of chemical and other control methods, when used with biological control agents.
3. Explain why a knowledge of organisms' life cycles is important in the correct application of the principles of biological control.
4. Explain what is meant by the term 'polyphagous' predator.

Answers

Exercise 23.1. Common uses of biological control agents

Results

1. (a) Black parasitized scales
    (b) parasite
    (c) whitefly.

2. (a) Pirate bug black and orangey-brown colour
    (b) predator
    (c) thrips
    (d) notify the Ministry.
    (e) *Macrolophus caliginosus* controlling whitefly.

3. (a) Orange quick-moving mite much larger than red spider mite.
    (b) predator
    (c) red spider mite
    (d) suitable examples, e.g. *Therodiplosis*, *Typhlodromus* and *Amblyseius* all control red spider mite; *Hypoaspis miles* controls sciarid fly; *Amblyseius* spp control thrips.

4. (a) tiny microscopic worm-like creatures
    (b) parasite
    (c) slugs
    (d) *Steinernema carpocapsae* and *Heterothabditis megidis* both control vine weevil.

5. (a) Presence of white and brown gungal mycelium decomposing whitefly
    (b) pathogen
    (c) whitefly
    (d) aphids and thrips.

6. (a) The bacteria cannot be seen. However the caterpillar appears black, deshevelled and limp
    (b) pathogen
    (c) caterpillars

(d) pheromone traps and *Trichogramma evanescens* parasitic wasp.

Conclusions

1. The use of a natural enemy as a predator, pathogen or parasite of a pest or disease.

2. Use blue sticky traps that are not attractive to *Encarsia*; introduce *Encarsia* early; use the dribble technique and ensure at least 21°C.

3. An organism that feeds on a living pest, eventually killing it. Mostly, only one pest is killed during the life cycle of the parasite. For example the wasp *Encarsia formosa* lays its eggs in the larvae (scale) stage of whitefly. As the wasp egg hatches the grub eats away at the protein-rich scale, killing the pest.

4. Anagrus atomus wasp controlling leaf-hopper or other example.

5. An organism that eats another, such as a fox eating a chicken. They may kill many pests during their life. For example the mite called *Phytoseiulus persimilis* hunts and kills red spider mite.

6. For example *Delphastus pusillus* predatory beetle controlling aphids.

7. Any two examples from Table 23.1.

8. Any four examples from Table 23.1 suitable in glasshouse situations.

9. Pressure from consumers for greener and cleaner food and produce together with pressure from the environmental lobby.

Exercise 23.2. Diagnosis of pest damage

Results

An appropriate table describing each pest and the symptoms of their damage should be produced.

Conclusions

1. Biting/chewing insects have mouth parts that rip apart and shred plant tissue. Sap sucking insects by comparison have a stylet with which they pierce plant tissue.

2. The lack of cell turgor more easily enables pest attack, and plants with insufficient nutrition are less resistant to disease.

3. Fungal diseases easily grow on honeydew left from sap-sucking insects or pentrate damaged tissue from biting/chewing insects.

Exercise 23.3. Monitoring pest levels

Results

A variety of pests are likely to be recorded including sciarid flies, thrips, aphids and whitefly. The blue trap attracts more thrips than the yellow.

Conclusions

1. Biological control will increase due to consumer pressure for 'natural' foods grown without chemicals and due to environmental pressure for greener and cleaner produce. Difficulties in use include: slow to establish and gain control; live organisms that sometimes die in transport; operators require knowledge of life cycle stages, pesticide and insecticides use restricted; correct environmental conditions are needed (e.g. light and humidity); may consume other natural enemies (e.g. *Macrolophus* eating *Phytoseiulus*); less successful on outdoor plants due to dispersal and colder temperature; and life forms are unpredictable and may be difficult to rear.

2. Suitable example e.g. whitefly control using *Encarsia formosa*. *Encarsia* should be applied when there is less than 1 whitefly/10 plants. The best control is achieved by a 'little and often' introduction of the wasp. This is sometimes called the dribble method. 1 wasp/plant at each stage of introduction; 5 introductions at 7–10 day intervals. The wasp is normally applied at 100 scales to the centre of a 100 plant block. It has been found that the wasp can fly the maximum distance from the centre of this block to the furthest plant (approx 7 m). The next application would be put in the intermediate position between previous

introduction sites. When applying the black scales, they should be in a sheltered position, usually low down on the plant and away from sunlight. Great care needs to be used with other chemical controls. For ornamentals apply 5–10 parasites/m², when density of whitefly is no more than 1 per 2 m². With young rooted plants it is important to introduce parasites early.

With fuchsia, pelargoniums and poinsettia control should be introduced when stock beds are planted. *Encarsia* will not move around or fly in temperatures less than 20°C, and are therefore not particularly efficient parasites at low temperatures. Whitefly easily multiply at low temperatures (below 12°C), therefore *Encarsia* is normally used from mid-April when temperatures are rising, however the glasshouse may be heated for just an hour to stimulate wasp activity.

3. Two examples from Table 23.1.

4. Monitoring and sampling is the basis of intelligence gathering in order to gain effective control before pest numbers become a problem. There are several techniques including regular plant walk inspections, pit-fall trapping and using sticky traps.

5. Blue traps attracts four times as many female thrips as yellow traps and therefore give earlier detection of pest problem.

6. Day-length, light radiation and temperature which affect insect behaviour.

7. Suitable control selected from Table 23.1.

Exercise 23.4. Predators and prey

Results

Refer to Table 23.2.

Conclusions

1. An organism that eats another, such as a fox eating a chicken. They may kill many pests during their life. For example the mite called *Phytoseiulus persimilis* hunts and kills red spider mite.

2. Chemical use should be kept to a minimum and approved for use with the biological agent being used.

3. Biological control requires that the horticulturalist has knowledge of pest life cycles so that action to break the life cycle can be taken. To take action against the adult, knowing that the larvae, pupae and eggs are still a threat, would not give control. Hence an integrated approach to breaking each stage of the life cycle should be followed. That may mean timely reapplications of controls to eliminate newly hatched offspring.

4. An organism that feeds on more than one type of prey.

List of useful suppliers

Most materials suggested in this publication are easily ordered locally through garden centres and chemists. The following companies have been found to be reasonable suppliers of some of the more difficult material to find.

Bunting Biological Control Ltd
Westwood Park
Little Horkesley
Colchester
Essex CO6 4BS
Tel: (01206) 271300
Fax: (01206) 272001

Biological control products.

Defenders Ltd
Occupation Road
Wye
Ashford
Kent TN25 5EN
Tel: (01233) 813121
Fax: (01233) 813633

Mail order biological control products.

Griffen Education
Griffen & George
Bishop Medow Road
Loughborough
Leicestershire LE11 5RG
Tel: (01509) 233344
Fax: (01509) 231893

Educational supplier, microscopic slides and general laboratory apparatus.

Philip Harris Education
Lynn Lane
Shenstone
Lichfield
Staffordshire WS14 OEE
Tel: (01543) 4822204
Fax: (01543) 480068

Educational supplier, microscopic slides.

Koppert UK Ltd
1 Wadhurst Business Park
Fairchrouch Lane
Wadhurst
East Sussex TN5 6PT
Tel: (01892) 784411
Fax: (01892) 782469

Biological control products.

Merckoquant (BDH)
Merck Ltd
Hunter Boulevard
Magna Park
Lutterworth
Leicester LE 17 4XN
Tel: 0800 223344
Fax: (01455) 558589

Educational supplier, laboratory apparatus and chemicals, tetrazolium salt, pH kits.

Bibliography

Adams, C.R., Bamford, K.M. and Early, M.P. (1995). *Principles of Horticulture*, 2nd edn. Heinemann.

Bagust, H. (1992). *The Gardener's Dictionary of Horticultural Terms*. Cassell.

Ball, R. and Archer, G. *Integrated Science Assignments – Communication and IT Skills*. Cambridge University Press.

Buczacki, S. and Harris, K. (1989). *Collins Guide to the Pests, Diseases and Disorders of Garden Plants*. Collins.

Clegg, C.J. and Mackean, D.G. (1994). *Advanced Biology. Principles & Applications*. John Murray.

Day, D. (ed). (1991). *Grower Digest 11, Biological Control – Protected Crops*. Grower Publications.

Finagin, J. and Ingram, N. (1988). *Biology for Life*. Thomas Nelson.

Food and Agriculture Organisation (1984). *Fertilizer and Plant Nutrition Guide*. Food and Agricultural Organisation of the United Nations.

Freeland, P.W. (1988). *Investigations for GCSE Biology*. Hodder & Stoughton.

Freeland, P.W. (1988). *Investigations for GCSE Biology – Teachers Book*. Hodder & Stoughton.

Freeland, P.W. (1991). *Habitats and the Environment: Investigations*. Hodder & Stoughton.

Gratwick, M. (ed). (1992). *Crop Pests in the UK, collected edition of MAFF leaflets*. Chapman & Hall.

Hessayon, D.G. (1991). *The Lawn Expert*. pbi Publications.

Hessayon, D.G. (1992). *The Tree & Shrub Expert*. pbi Publications.

Hodgson, J.M. (1985). *Soil Survey Field Handbook,* second edition. Soil Survey of England & Wales. Technical Monograph No.5. Bartholomew Press.

Johnson, A.T. and Smith, H.A. (1986). *Plant Names Simplified. Their Pronunciation, Derivation & Meaning*. Hamlyn.

Koppert, B.V. (1993). *Koppert Bio-journal*. Koppert.

Koppert, B.V. (1994). *Koppert Products – with directions for use*. Koppert.

Mackean, D.G. (1983). *Experimental Work in Biology*. John Murray.

Mackean, D.G. (1989). *GCSE Biology*. John Murray.

Ministry of Agriculture Fisheries & Food (1983). *Lime in Horticulture*. ADAS leaflet 518. HMSO.

Phillips, W.D. and Chilton, T.J. (1994). *A Level Biology*, revised edition. Oxford University Press.

Press, B. (1993). *Bob Press's Field Guide to the Wild Flowers of Britain & Europe*. New Holland.

Roberts, M.B.V. (1980). *Biology – A Functional Approach – Students Manual*. Thomas Nelson.

Rose, F. (1981). *The Wild Flower Key to Britain and N.W. Europe*. Warne.

Rouan, C. and Rouan, B. (1987). *Basic Biology Questions for GCSE*. Bath Press.

Rowland, M. (1992). *Biology University of Bath Science 16–19*. Thomas Nelson.

Sampson, C. (1995). Whitefly predator debut. *Grower*, February 23.

Scott, M. (1984). *Efford Sandbeds*. ADAS Leaflet No. 847., MAFF. HMSO.

Simpkins, J. and Williams, J.I. (1989). *Advanced Biology*, third edition. Unwin Hyman.

Soper, R. (Ed). (1990). *Biological Science 1 & 2*. Cambridge University Press.

Whitehead, R. (Ed). (2007). *The UK Pesticide Guide. British Crop Protection council. CAB International*. Cambridge University Press.

Woodward, I. (1989). Plants' Water and Climate (Inside Science No.18). *New Scientist* (18 February).

Wye College (1995). *Wye College Biological Control Handbook*. Wye College Press.

Index